Inventing the 21st Century

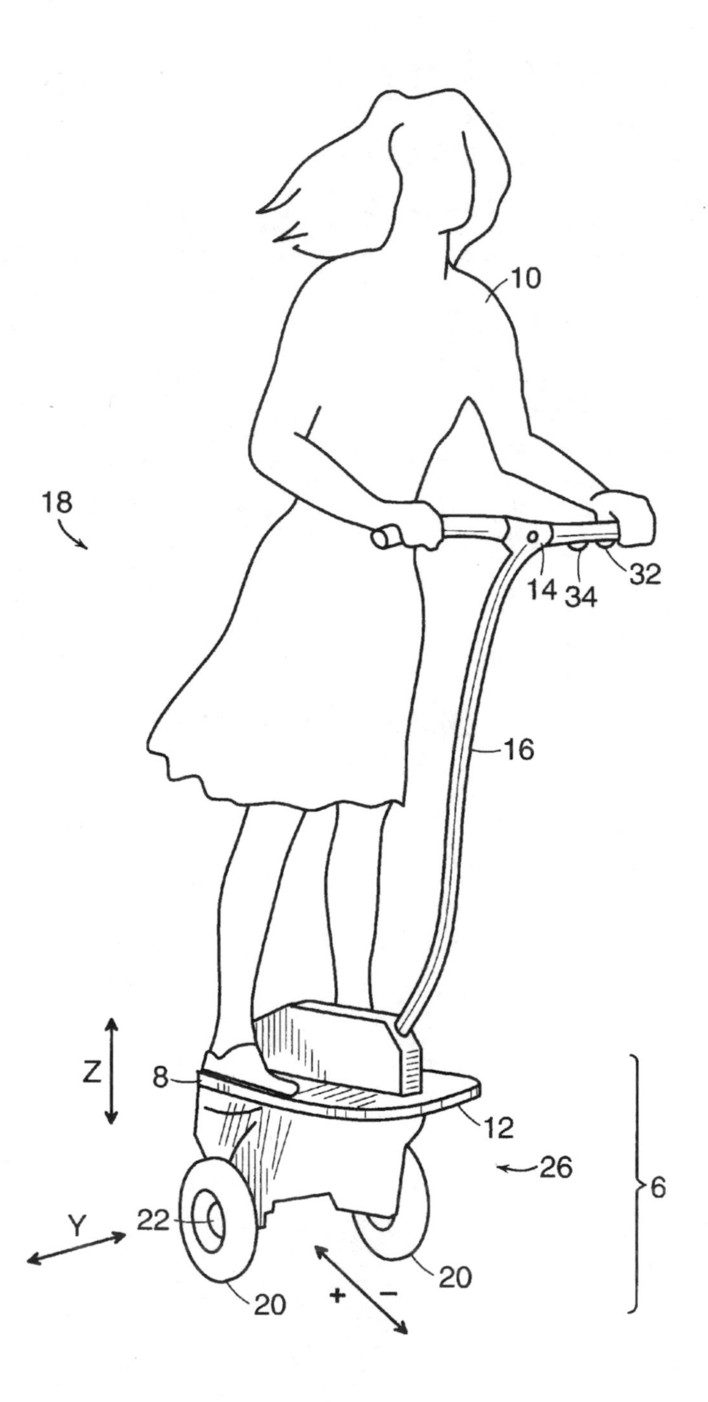
10
18
14
34
32
16
Z
8
12
26
6
Y
22
20
20
+
–

Inventing the 21st Century

Stephen van Dulken

Contents

Introduction

THIS BOOK CONTAINS STORIES about inventions published, or at least popularised, in the first decade of this century, which I hope will interest, entertain and perhaps inspire you. Not all were successful, as this is a field where many are called but few are chosen. Nor would all be regarded as positive, if only because so many inventions tend to isolate us from others (think of mobile phones, personal stereos, cars, automatic checkouts . . .).

Some inventions more or less chose themselves for inclusion in this book as they have become so iconic. These are combinations of electronic gadgets and the ubiquitous software which is now essential in so many products and services. We are only a decade into the new century but already this information age has exerted a huge influence on our everyday activities, often without our realising it. In many cases we are working (or living?) smarter rather than harder. Also included are apparently mundane inventions such as coffee-cup holders which are also easily taken for granted. In every case, someone had to design them in the first place. Other inventions were chosen because of an interesting story behind them, often coming to my attention on the television show *Dragons' Den* (or *Shark Tank*, its equivalent in the USA), in which applicants for finance are cross-examined by a panel of venture capitalists.

I have worked as a patent librarian, a rare speciality, since 1987. It is a mysterious area where technology, law and business meet. In recent years I have also been involved with business as a subject. I blog, write books, give talks or conduct workshops, run priced searches for clients, help enquirers and so on. Trade marks and designs and to some extent copyright are involved as well as patents. Uniquely, in Britain patents are held in the national library, but then the British Library inherited the Patent Office Library, founded in 1855, and its collection of over 55 million patents. Other countries took a different approach and kept the collection of patents with the patent office itself.

My work has given me an interesting viewpoint on inventions. I am in one sense involved in the trade, but in another very real sense I am not, as I do not share in the money won or lost in innovation, and am at best only a link in the chain of events needed to bring a new product to market. As a librarian and searcher I enjoy looking at the actual patent specification as it often makes explicit what is otherwise hidden in vague press releases. Technical knowledge, however, is often needed to understand all the details.

I become particularly involved with private inventors. As our discussions are confidential, I cannot give details, but it will surprise nobody that the Internet and mobile phones are very popular areas. They probably represent half of the enquiries I handle. Most of the inventors I meet are thoughtful people who have worked out their needs for finance, market research and business plans. It tends to be, however, the rare exceptions that stay in the memory. A wide portfolio of skills is expected of inventors, and any individual who feels unable to present their inventions adequately must be under a huge strain. Major companies have the great advantage of specialist departments that will search to see if an invention appears to be new, patent it, and then market it.

If I think an idea presented to me has been tried before, or is unlikely to find a market, I sometimes ask whether the inventor would say yes if offered a million pounds for all rights. The answer has always been no. One inventor nearly burst a blood vessel, saying that any such offer would 'cheat' him. I know that I am not a risk taker and would happily accept such an offer and live off the interest, with no desire to risk trying for more, or put in extra hours. I'd simply put my feet up and think of the next invention.

Entrepreneurs, however, are made of sterner stuff. These inventors can certainly be described as risk takers. I think the highest amount an inventor predicted in sales for his invention was £700 million, while I once spoke to someone whose first words were that I was looking at a future millionaire. It turned out that he thought that he could legally monopolise a certain type of business on the Internet (he couldn't).

Some inventors are either very naïve or very suspicious. Those in the first camp often join the second after their first invention, claiming that their idea was stolen. Some inventors will only reluctantly admit that an invention is, for example, sports related, while others are only too keen to tell me all about it. After all, they might say, all their friends liked it as well. This means that confidentiality has been lost (this is preserved if non-disclosure agreements are used, or in client relationships). Contracts are often not thought of, like the man who told me that he had spent a year developing his invention with a company on the basis of a handshake and without contracts covering finance, responsibilities and liabilities. 'But I trust them,' he told me.

What is needed for an invention to be successful? There must be a market for it of a reasonable size, either one that already exists or one that can be created by a totally new concept. It needs to be capable of being produced at a fair price for what it is, though if it is really cool and effective some will buy at the higher price that usually prevails at the beginning of big sales of a new product. The marketing also needs to be appropriate and well thought out, including, ideally, a good trade mark.

What skills does an inventor need? A knowledge of the industry is not just helpful but crucial. 'I have an idea that will revolutionise the baking industry,' one man eagerly said to me. Untypically, I asked, 'How much do you know about that industry?' 'Nothing,' he admitted. That is not unusual. Very often people say to me that they want to go into an industry, or combine products from two industries, when they know nothing of the industry and indeed very often nothing about finance, marketing, and so on – they just have the idea. This rarely seems a problem to them. Often they have no money either, and are indignant at the thought that those who do won't fall over themselves to give them financial backing. 'But it will sell in the millions,' they assure me. *Prove it*, I think.

It is vital that inventors try to understand the point of view of the person – retailer, manufacturer, backer – on the other side of the negotiating table. Likely questions should be thought of, facts and figures should be memorised. The person on the other side of the table is rarely interested in how clever the idea is, just whether it will make money. As a reality check on expectations, I often ask, 'If a product sells for £10, how much does it cost to make, and how much can be expected in royalties?' The reply is often over-optimistic, particularly concerning royalties, which in practice would commonly be a maximum of 40p per unit; the manufacturing cost would be in the region of £2.50.

Even if these barriers are overcome, the invention must be described in a patent specification so that someone 'skilled in the art' can reconstruct it. Each patent has what are called 'claims' which set out the protection granted. These claims must be carefully worded so as to maximise what is asked for by the applicant. You don't ask for something made 'of steel', but made 'of a suitable material', for example. The invention has to be new, but does not have to work, hence the glee of the inventor who pointed out to me that the UK 1977 Patents Act did not specifically ban perpetual-motion patents.

If a patent is granted it won't earn money unless what is patented is manufactured, licensed out, or sold. Often inventors think that a patent automatically prevents competition. Patent infringement is a civil and not a criminal offence so you have to fight a court action to stop infringers, and if the market for a product is really as lucrative as many inventors suggest then large companies will jump in to compete.

I often advise clients on searching to see if their idea is new. They may do this themselves at work stations in the library using free Web databases, but I also often carry out priced searches in a backroom. I combine my searching skills with the specialist knowledge of the client, though if they have no knowledge of the industry the specialist terms have to be found somehow. Many do not realise that searches are carried out using words and classifications, not by examining drawings, no matter how detailed.

Searching can be more complicated than it seems. One son of a client, convinced that I was trying to cheat his mother, complained, 'It's not rocket science!' No, it's often more complicated. An invention may be easy to describe but difficult to find in a database. For example clients often say the invention should be portable. Suppose the searcher adds the word 'light' to convey the idea of a lack of weight. Yet that word could be taken to mean illumination. Synonyms for 'sensor' abound. Queueing is waiting in line, but is also a computer term. Clothing can be described as clothes, apparel, raiments or a specific garment (or part of it, like a sleeve). Many inventors are content with a quick search using free sources, but a couple of days looking for and reading results is recommended as a minimum before investing a lot of time and money in an idea that has often been done before.

The perfect search would find everything relevant but exclude anything irrelevant. This is impossible, and nobody will promise it. You cannot prove a negative. This is not just because patent applications take 18 months to publish, so a relevant filing may be awaiting publication and hence detection. The aim is to have good 'recall' of relevant material but to try to avoid less relevant material, ideally by sorting the results into sets of descending relevance. The database will occasionally have misspellings or will describe an invention using contorted language.

I usually write a complicated search strategy using what is called Boolean logic (using the terms *and*, *or*, *not* to link words or classes, or sets of data). This means that data can be manipulated to create a 'limited set' of relevant results. These are then subdivided into unique sets so that the most likely hits come first. It is only at this stage that patent titles are actually seen, as before then only the number of hits in reply to each query are seen. It is then up to the clients and their legal advisers to decide which ones are relevant, and to ask for more details.

I often receive enquiries from the media, mostly either television companies or newspapers. It is frustrating if they then go on to get it wrong in their reports, but at least they asked. My blog fights a constant battle against sloppy wording such as 'Smith has a patent' when they mean 'Smith has *applied* for a patent'. Perhaps half the enquiries from the media are attempts to present inventions or inventors as eccentric. I wonder how they would react if someone ridiculed their own profession in that way.

To conclude, I am frequently asked about my books by people I meet. The two most popular questions are:

How long did it take you to write it? (No idea, I didn't count the hours.)

Did you do the drawings? (No, draughtsmen, or occasionally the inventor, did.)

A close third is, 'Does the British Library have a copy of every book ever published?'

No, we don't. But we do have a very large number.

Note: Many trade names mentioned in this book are registered trade marks or those asserting the use of an unregistered trade mark. Registered trade marks are commonly indicated by a capital letter for the first letter of the trade name.

The rise of e-books

The days of paper books are perhaps doomed now that it is possible to carry around a library in a personal e-book reader. The basic idea is that books can be downloaded onto an electronic device the size of a paperback, hence achieving the prize marketing goal of requiring the purchaser of a device to spend a lot on aftersales. In fact, as there is coyness about revealing sales figures for the reading devices, it is possible that hype rather than sales is driving the concept forward.

The Sony Reader was launched in November 2006. The original model sold in the UK for £199 and had 200MB of memory, enough for 160 average-length books. A CD containing 100 classic texts was included in the price. Power demand is traditionally the problem with such devices. Sony's device only uses power when pages are turned, and the battery life is enough for 1,680 page turns. There is no glare, and it can be read in bright light. Only a thin display screen is needed, too.

The low power demand is due to the fact that the wording within the screen is based on electronic ink, or e-ink. This technology is owned by E-Ink Technologies of Cambridge, Massachusetts. The basic idea uses silicon 'rubber stamps' to print tiny computer circuits onto the surface. The electronic ink consists of millions of tiny capsules, filled with a dark dye, floating in a substance similar to vegetable oil. The capsules contain negatively charged white chips that move either up or down within the capsules in response to a positive charge applied to the medium's surface. That produces the words. It has been suggested that this technology will replace LCDs for many applications.

A problem, though, is deterioration in the light quality. This is faster in static areas, where the image is constant (such as where page numbers are given, or a running title at the top of the screen), and produces 'burn-in', which screen savers on PC monitors are designed to prevent. Sony has applied for a patent for an invention by which corrections are automatically made to compensate for this deterioration (see the illustration on p.12).

One problem that will have to be resolved is electronic publishing rights. Many authors and agents maintain that, as contracts for older books do not mention such rights, they reside with the author. Would-be e-book publishers, on the other hand, argue that phrases such as 'in book form' or clauses that prohibit 'competitive editions' prevent authors from publishing e-books through other parties. This kind of feud has occurred before over formats such as electronic games or hand-held devices, where agreements for rights were specified long before such formats were conceived.

There are competing products to the Sony Reader, which include Amazon's Kindle, developed by their start-up subsidiary Lab126. This company had said that it was working on a 'groundbreaking, highly integrated consumer product', and speculation at the time was that this would be a rival to the iPod or iTunes. Instead, Amazon came up with an e-book reader. The Kindle was launched in the United States in November 2007. The first run sold out in five and a half hours, and new stocks were not available until the following April.

Also using e-ink, it has a huge advantage over rival products in that books can be downloaded directly from the Amazon website. Sales on Christmas Day 2009 – admittedly not, perhaps, a day for big sales anyway – were said to be higher for e-books than for paper versions. Over 400,000 Kindle-ready books are available, and 3,500 books can be held in a single Kindle DX model.

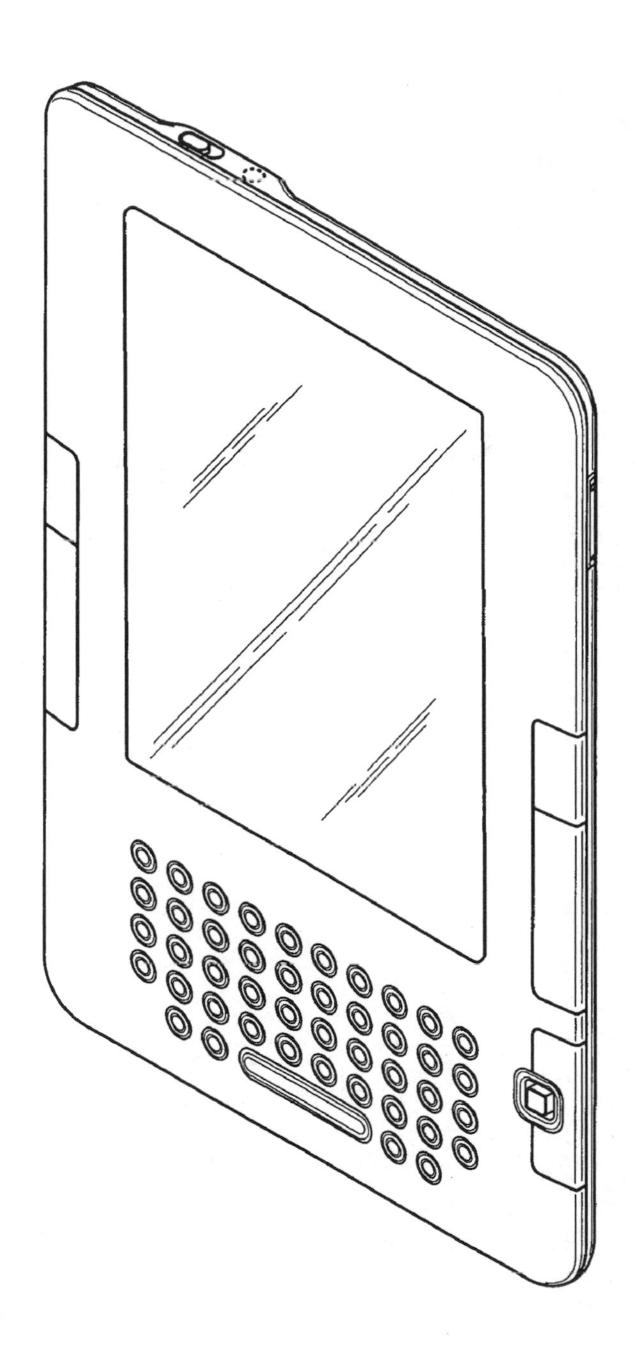

Electronic reader device,
US D601559, published 6 October 2009

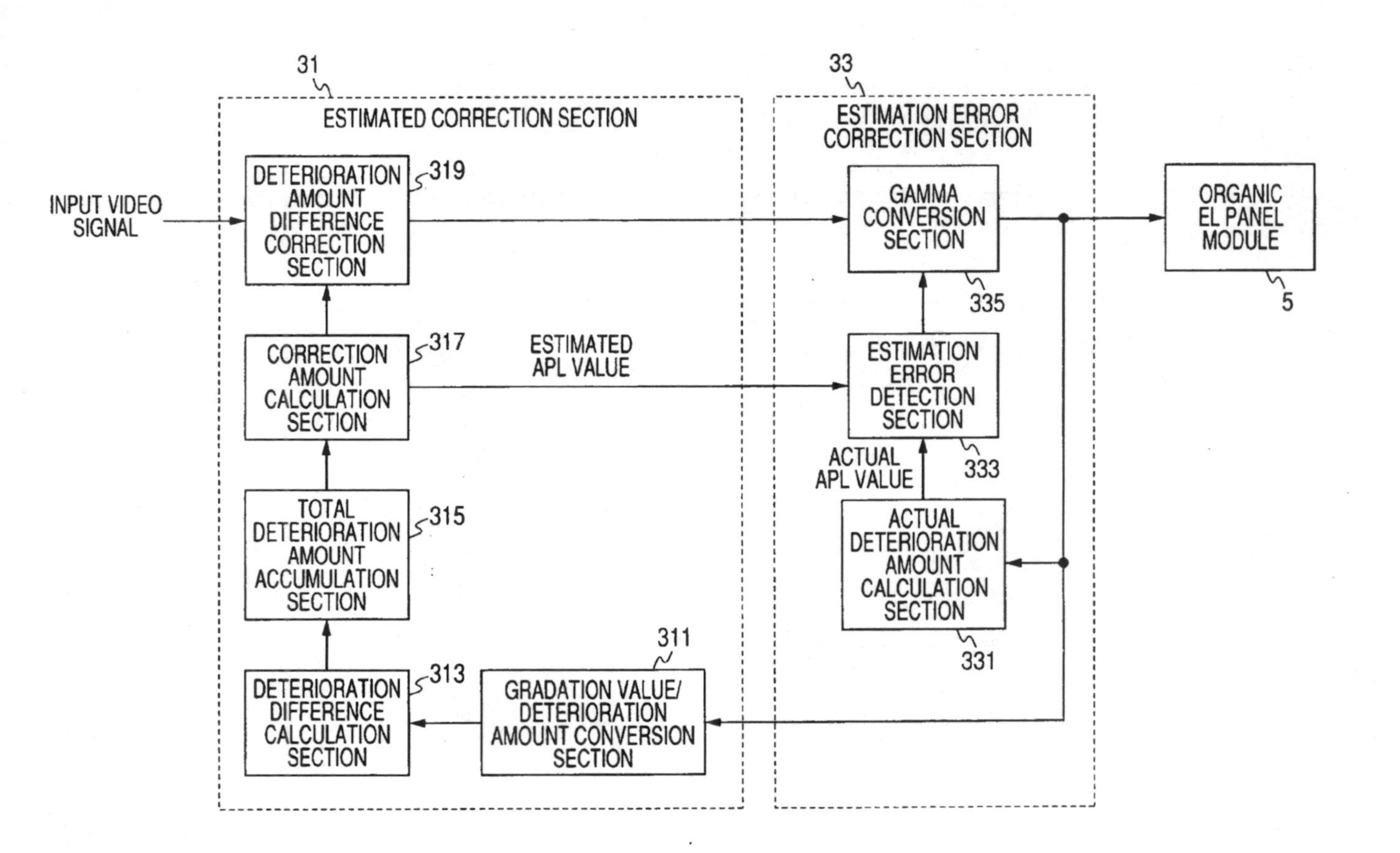
31
ESTIMATED CORRECTION SECTION
33
ESTIMATION ERROR CORRECTION SECTION
INPUT VIDEO SIGNAL
DETERIORATION AMOUNT DIFFERENCE CORRECTION SECTION
319
GAMMA CONVERSION SECTION
335
ORGANIC EL PANEL MODULE
5
CORRECTION AMOUNT CALCULATION SECTION
317
ESTIMATED APL VALUE
ESTIMATION ERROR DETECTION SECTION
333
ACTUAL APL VALUE
TOTAL DETERIORATION AMOUNT ACCUMULATION SECTION
315
ACTUAL DETERIORATION AMOUNT CALCULATION SECTION
331
311
DETERIORATION DIFFERENCE CALCULATION SECTION
313
GRADATION VALUE/ DETERIORATION AMOUNT CONVERSION SECTION

In January 2009 Amazon, using nine named designers, filed a design patent US D601559 for the look of an 'Electronic reader device'. In January 2010 Apple unveiled their customary secret-until-Steve-Jobs-announces-it new product: the iPad. It had been expected that this would be a tablet, that is a 'notebook' computer without a mouse or keyboard. Replacing the mouse is easy, by using the touch screen instead, but the keyboard migrates to the screen. The bright, clear screen is really meant for those who want to read books or watch films (the screen is somewhat bigger than the other e-book readers), rather than create their own material or do a lot of typing. It *is* more versatile than a dedicated e-book reader, which should help sales. A price war between Sony, Amazon and Apple is likely, if only in the downloaded titles.

Those who like the way Apple's products work will love the iPad. They will see it as slick and easy to use. It incorporates much technology already used in the iPod and iPhone, but some ideas used in the iPad are new, and are included in their US 2008/259039, 'Method, system, and graphical user interface for selecting a soft keyboard'.

Light-emitting display device, US 2007/236431, published 11 October 2007

A smart wheelchair

Back in 2000 Mike Spindle was at Luton airport with his family and noticed a stylishly dressed teenager in a decidedly unstylish wheelchair. The contrast jarred and he began to sketch designs on the back of his boarding pass.

His initial thought was not to look at existing wheelchair designs but to create a completely new one. He began to consider how wheelchairs could be made more useful for their users. Spindle ran his own small engineering company, D.T. Clayton (Toolmakers) Limited, so he had both the skills and the tools he needed.

After years of experimenting, in 2006 the Radlett-based inventor applied for a trade mark for the product, Trekinetic, and a patent. There was a lot of detail: 42 pages of text and 16 pages of drawings. Patent specifications have to describe an invention in enough detail for an expert in the same field to be able to reconstruct it.

Inventors often explain in detail the problems their inventions are meant to solve. In this case, Spindle explains in his patent application that conventional wheelchairs have four wheels, with the user moving the wheelchair by propelling the rear pair of wheels, while the front pair of castors enable turning. This design makes its use less than perfect on uneven or uphill surfaces. Some alternative designs reverse this concept (the front wheels are the drivers), which can make self-propulsion easier for many.

Even so, if only three of the four wheels are in contact with the surface, instability occurs. The high centre of gravity does not help. Conventional wheelchairs also have an understandable tendency to roll down slopes unless the brakes are on. Finally, it can be difficult to assemble a collapsible wheelchair, or one in parts.

A number of such problems were solved in the redesign. A monocoque carbon-fibre seat (whose skin absorbs the stresses), was used to which the other components, including two large front wheels and a single, smaller rear wheel, were fitted. Conventional designs would use a tubular steel frame across which canvas is stretched to form the seat. The Trekinetic can easily fold up, with only the wheels needing to be detached, and screws rather than welding are used, so that it is easy to repair if necessary.

The biggest problem Spindle faced was getting the new machine to run in a straight line. Three wheelers do not naturally do this, but the problem was eventually solved by using a spring-loaded catch inside the rear castor. Spindle also designed a system of weight transfer where the user's weight can be transferred either forwards or backwards automatically by the occupant, in-chair. This allows the user to ascend high kerbs without needing to do 'wheelies' as they would in a conventional wheelchair. He used a rear shock absorber that allows the seat to recline (this position is shown in the drawing). The user simply has to lean back.

The Trekinetic is also capable of faster speeds than conventional wheelchairs. The drum brakes ensure a swift halt if necessary. The seating position is better supported than previously, so there is less unwelcome pressure, and the cushions are designed to minimise what pressure there is.

Trekinetic tyres are wider than on most wheelchairs and are treaded to assist movement across rough ground. Wheel camber is used – the wheels are angled to form an inverted V – so that the track is wider at the bottom, giving more stability at speed. With such a system, narrow doorways could present a problem, so the Varicam arrangement

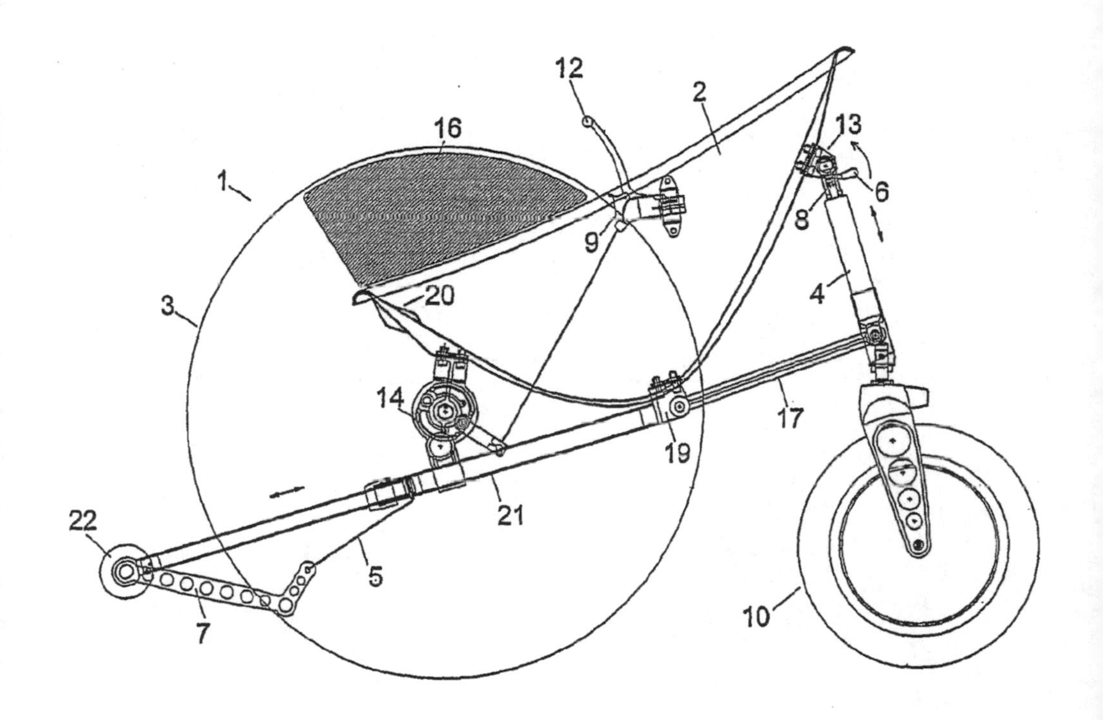

BACKREST RECLINED

Improved wheelchairs and wheeled vehicles, WO 2007/091022, published 16 August 2007

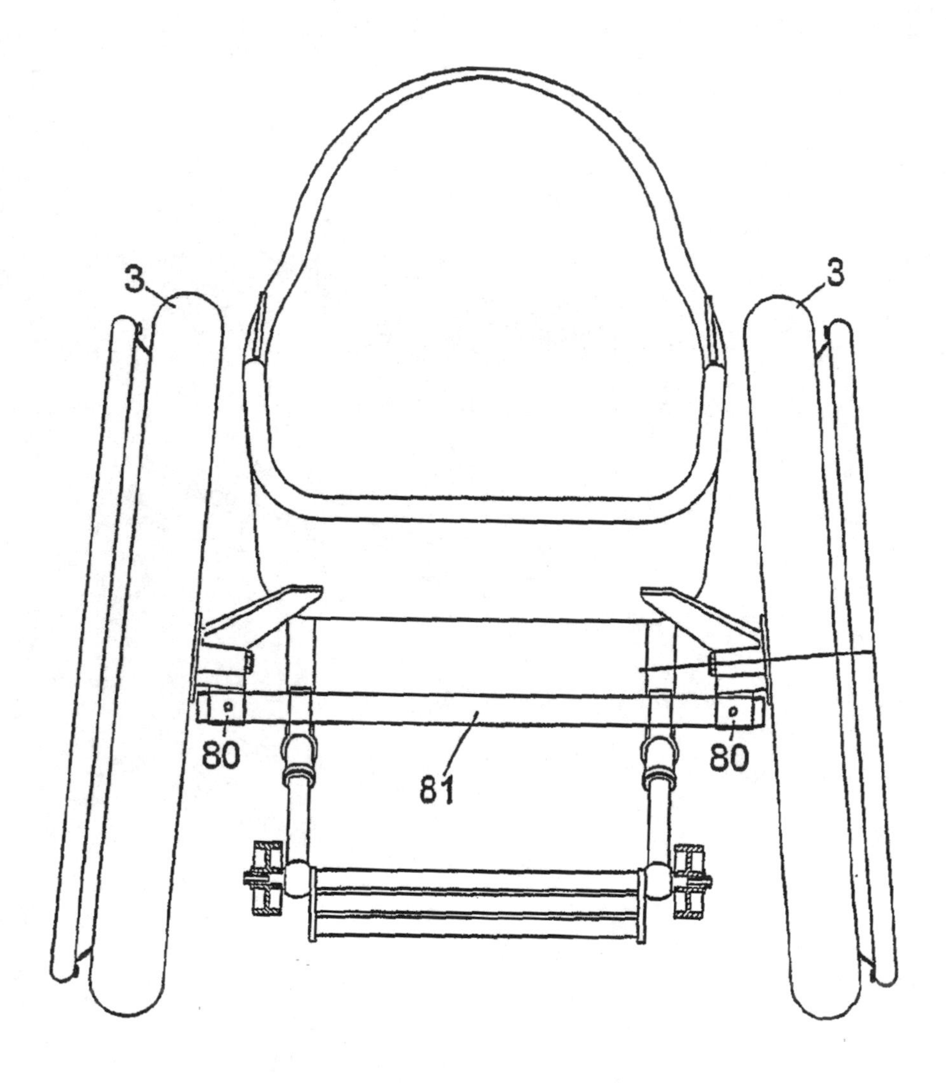

Improved wheelchairs and wheeled vehicles, WO 2007/091022, published 16 August 2007

was invented: a threaded shaft under the seat that can be used to pull the wheels upright for home use, reducing the maximum width between the wheels from 870mm to 710mm.

Although costing about six times as much as a typical NHS model, this first all-terrain wheelchair has been enthusiastically received. Not only do the users love it ('a new world opens'), but onlookers are also interested and often gather round – a distinctly unusual reaction to wheelchairs – with remarks such as, 'Hey, cool wheels.'

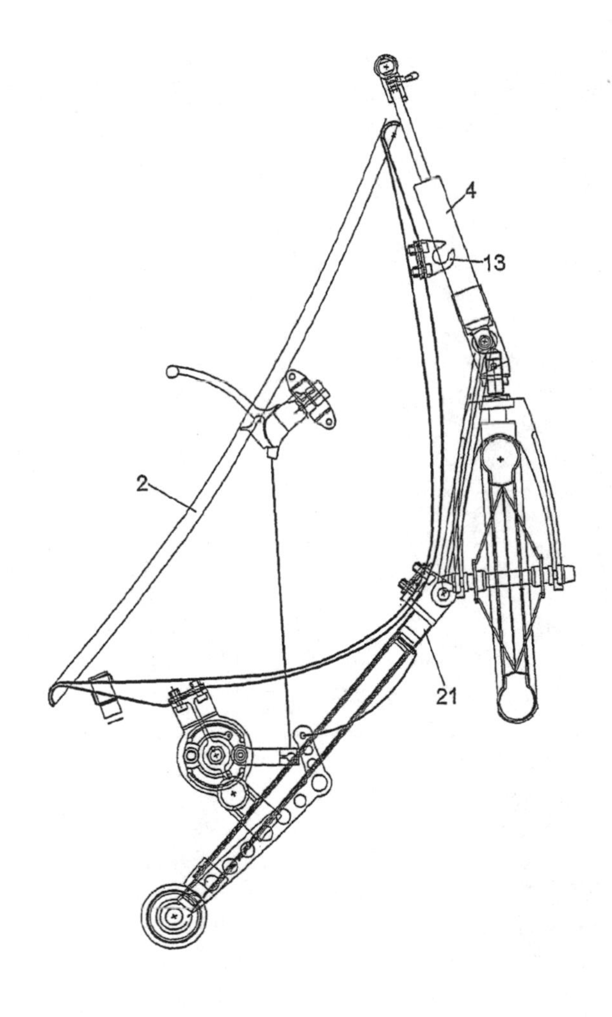

The 'bladeless' fan

The Dyson Air Multiplier fan was launched on 13 October 2009. Instead of visible blades it consists of a hoop above a base. The blades are, in fact, present, but are hidden inside the base with the motor.

Dyson's press release explains, 'Air is accelerated through an annular aperture set within the loop amplifier. This creates a jet of air which passes over an aerofoil-shaped ramp that channels its direction.' This produces a flow of 405 litres of air every second. Surrounding air is drawn into the airflow (this is called inducement and entrainment). The flow of air moves over the curved surface, which brings in more air to create a draught, in much the same way as air flows over an aircraft wing. This creates the 'multiplier' effect – the company claims that fifteen times more air is expelled than is pushed out by its motor.

Dyson's fluid dynamics engineers ran hundreds of simulations to measure and map the airflow so as to maximise the device's efficiency. The fan is powered by a brushless motor and air speed can be precisely adjusted with a dimmer-switch-type control. Conventional bladed fans usually have only two or three settings.

No visible blades mean no need for a grille. This, in turn, means that the fan is simple and safe to clean, with no danger to fingers. Because the motor is at the base, the fan is not top heavy and can be tilted with a touch. Nor is there the usual buffeting of conventional fans. When asked whether removing key elements of established designs meant that he liked simplicity, Sir James Dyson said, 'We removed bags [from the company's range of vacuum cleaners] because they clog and fan blades because they chop the air.'

The new fan, which is considerably more expensive than conventional models, is made from acrylontride butadiene styrene. This is a tough thermoplastic typically used to make car bumpers, crash helmets and golf-club heads. It is also used in Dyson's vacuum cleaners and hand dryers. All three products, of course, involve the use of airflow, and both the fan and the hand dryer require the air to gush out of thin slits. This is not a coincidence. Sir James has said when the company was developing the hand dryer it was noticed that it was drawing in a great deal of air from its surroundings. They wondered how this effect might be put to use. An air-moving device with no propeller or fan blade was thought of, and three years of development and a year of testing followed.

A Japanese patent specification has, however, been listed as prior art by the European Patent Office. It is not in force and therefore there is no question of infringement, only a limitation as to what the Dyson patent can claim as new. JP 56-167897A is by Tokyo Shibaura Electric and was applied for in 1980. It describes a desk fan which also has the hoop design and where air is discharged into the hoop, from which it is forced through narrow slits.

According to Gill Smith, head of Dyson's patent department, 'The difference is all in the technology. We've spent many years developing the Coanda surface. The Japanese version does not have this feature.' A Coanda surface is the aerofoil ramp over which the air is pushed out of the ring of the fan. Because of the angle of the Coanda surface, surrounding air is sucked into the flow, creating a smooth and powerful blast of air.

Smith 'absolutely expected' Dyson to be granted a patent for its Air Multiplier.

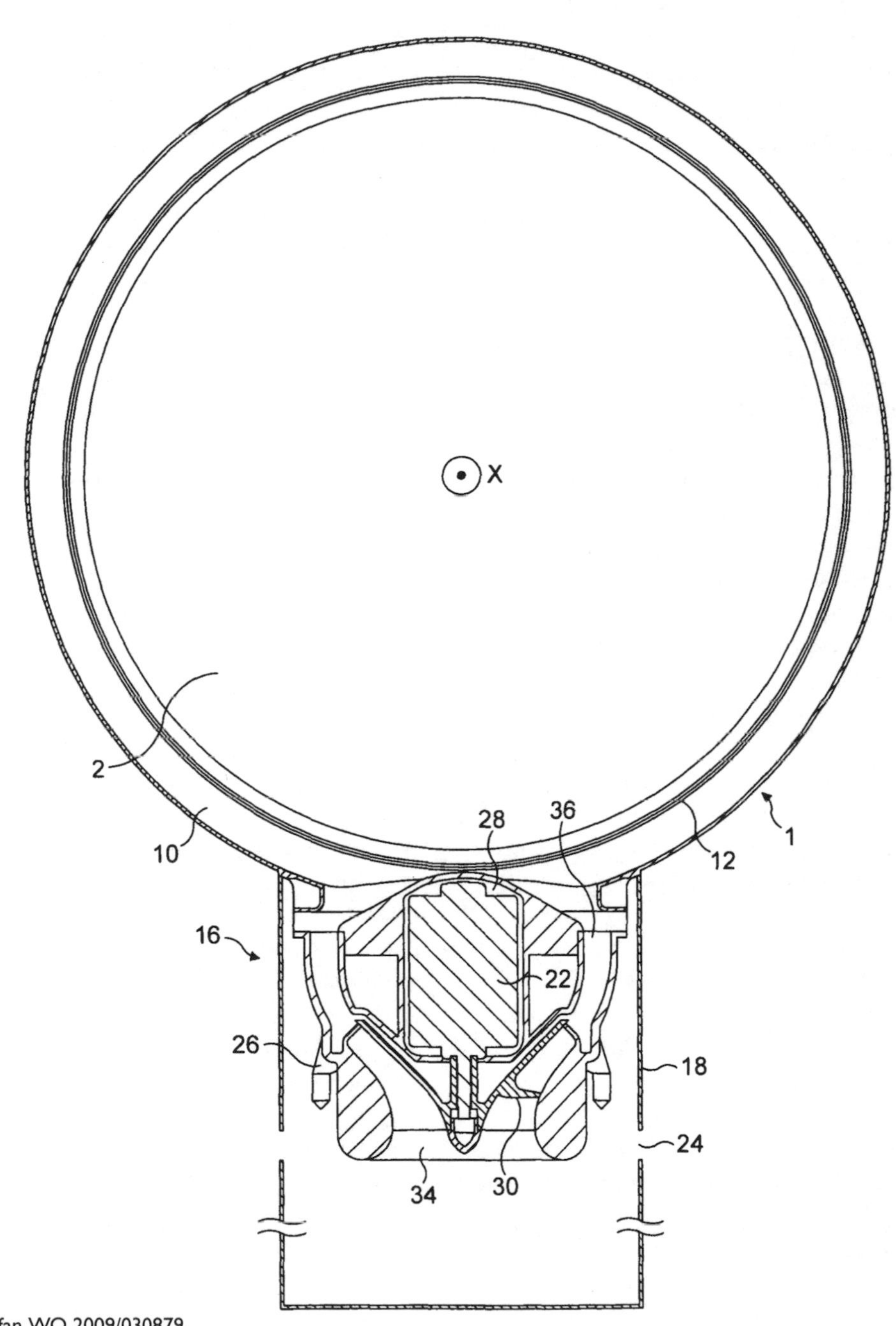

A fan, WO 2009/030879, published 12 March 2009

Tubular containers for confectionery

Packaging involves many innovations to ensure a cheap yet reliable product and here is one targeted primarily at the children's confectionery market.

The inventor, Mark Sheahan, has a background in the computer industry, having been a computer mainframe project manager at 21 overseeing 60 staff. After taking a variety of other jobs, including a stint as a croupier, he came more or less by chance into inventing about 15 years ago and found that he enjoyed the problem solving involved. He believes in keeping control over the rights to his inventions with licensing contracts for specific uses, rather than selling them. He is chairman of the Croydon Round Table of Inventors and is the British Library's Inventor in Residence, where he has met over 200 inventors in one-to-one discussions.

He had success with his earlier circular Squeeze Open containers, squeezing the flared sides of which causes the top to rise up and open. His new invention is also for an opening method.

Traditionally, containers of this type have a loose top which is prised open, like a tube of Smarties. Sheahan's is made of polypropylene, which gives it strength, and the whole top fits snugly inside the container. It is designed with a weakness at two points so that pressing on either the hinge side or the opposite side causes an outward deformation and a 'pop' sound as it opens. A holographic tag using 'wallpaper' (a repeated design) covers the non-hinge side to suggest the correct pressing point but also to show that it has not been tampered with before purchase and to avoid counterfeiting. Its shiny design is also attractive to children.

As a safety feature, there are tiny holes on the top, placed above the plastic rim, to avoid complete blockage of the airway if it is swallowed. The lid is not easily detached, reducing litter. The base of the tube has an internal design which is revealed once the contents have been eaten and the child looks down the tube. These are collectable as the designs will vary, prompting the name of the invention: Popi ('pop eye').

The top is injection moulded because of the complexity of the design, while the base is made by extrusion – a cheaper method using a cheaper plastic. A small bump or 'locator' on the underside of the lid enables accurate placing of the hologram during production.

The use of clear plastic means that the contents can be seen, unlike in cardboard containers. A possible future development is a square-sided container that will not roll off a flat surface. At present the tube is extruded from high-density polyethylene, though a cheaper blow-moulded version made entirely from polypropylene is planned, which will simplify both the manufacturing process and recycling after use.

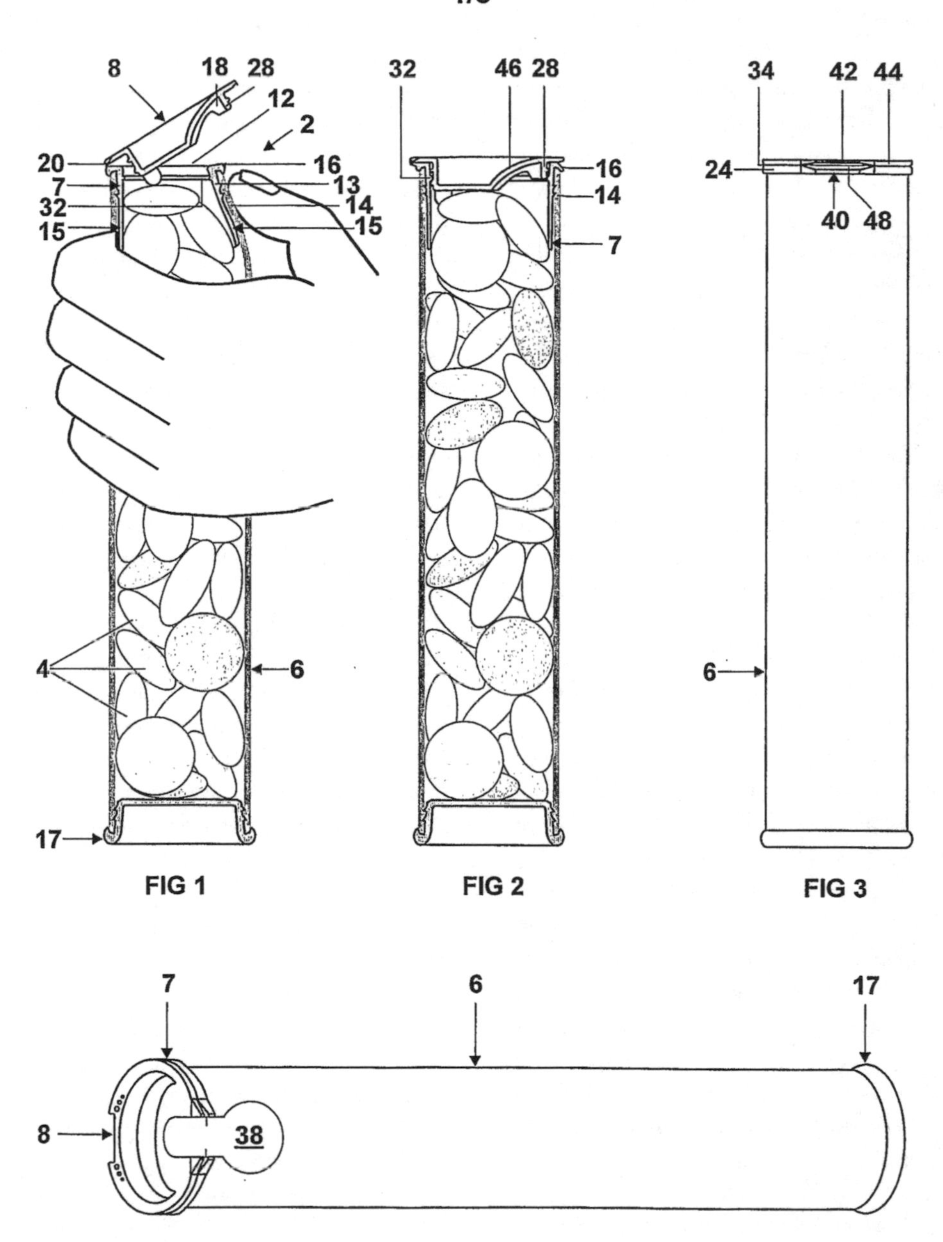

Tubular container, WO 2007/039711, published 12 April 2007

The iPod MP3 player

The iPod has been around for quite a while, of course, but made headlines in September 2008 when Apple admitted that someone else had thought up the original idea.

A surprising amount of work was involved in providing what, for millions, has become their favoured method of listening to music. Most reference sources credit German research organisation Fraunhofer Gesellschaft with the invention of a workable, good sound quality method of downloading music, for which a patent was filed in 1989 and which appeared in English as US 5579430 'Digital encoding process'.

The idea was to improve the sound-quality from previous attempts, particularly, the patent says, in WO 88/01811. That invention is for a data-compression algorithm which results in sound quality which is acceptable to most listeners (CD-ROM quality, says the specification) but which is not high fidelity. It does this by reducing the accuracy of the reproduction for notes beyond the hearing range of most people, or even discarding selected notes altogether. This is called perceptual coding, and JPEGs work in a similar fashion to save space when compressing images.

Fraunhofer launched the product in 1994. The compact size of the files enabled easy downloading and sharing and the first large 'peer-to-peer' music filesharing network, Napster, was launched in 1999. This, of course, led to accusations of copyright infringement and Napster's unlimited free service was shut down in 2001. Fraunhofer was estimated to have made ?100 million in licence fees for the algorithm in 2005 alone.

Apple decided to make a digital music player as existing models were 'big and clunky or small and useless', with user interfaces that were 'unbelievably awful'. The product was developed in less than a year and unveiled on 23 October 2001. Steve Jobs, CEO of Apple, said that the 5GB hard drive would put 'a thousand songs in your pocket'.

The name iPod was inspired by a line in the 1968 film *2001: A Space Odyssey* – 'Open the pod-bay door, Hal!' – spoken by an astronaut to an out-of-control computer. The product casing contains no screw holes or other crevices where dirt might get in, and its design soon became iconic. One black and white advertisement showed simply a silhouette of someone using an iPod and the familiar partly eaten apple below it – not a word or number in sight, but an instantly perceived reminder to the audience of both the product and the company.

It is probable that there are a number of patent specifications involved. I have tentatively identified two, both by Jeffrey Robbin and David Heller of Apple, US 2003/079038 and US 2003/167318. The first graphic shows the main drawing from the first of these, which is for downloading to a media player.

Sales have been phenomenal and the iPod continues to dominate the market, with over 220 million sold as of September 2009. It hasn't all been plain sailing, however. Burst.com approached Apple in 2004, alleging that Apple were using Burst.com's patented audio downloading technology and demanding a licensing fee for the same. After failing to agree on the sums involved, Apple sued Burst.com to try to invalidate its patents and Burst.com countersued, alleging patent infringement and claiming $500 million in damages.

In order to get themselves out of a hole, Apple required evidence of the existence of the technology prior to Burst.com's patent, and hence of the MP3 player itself. They apparently

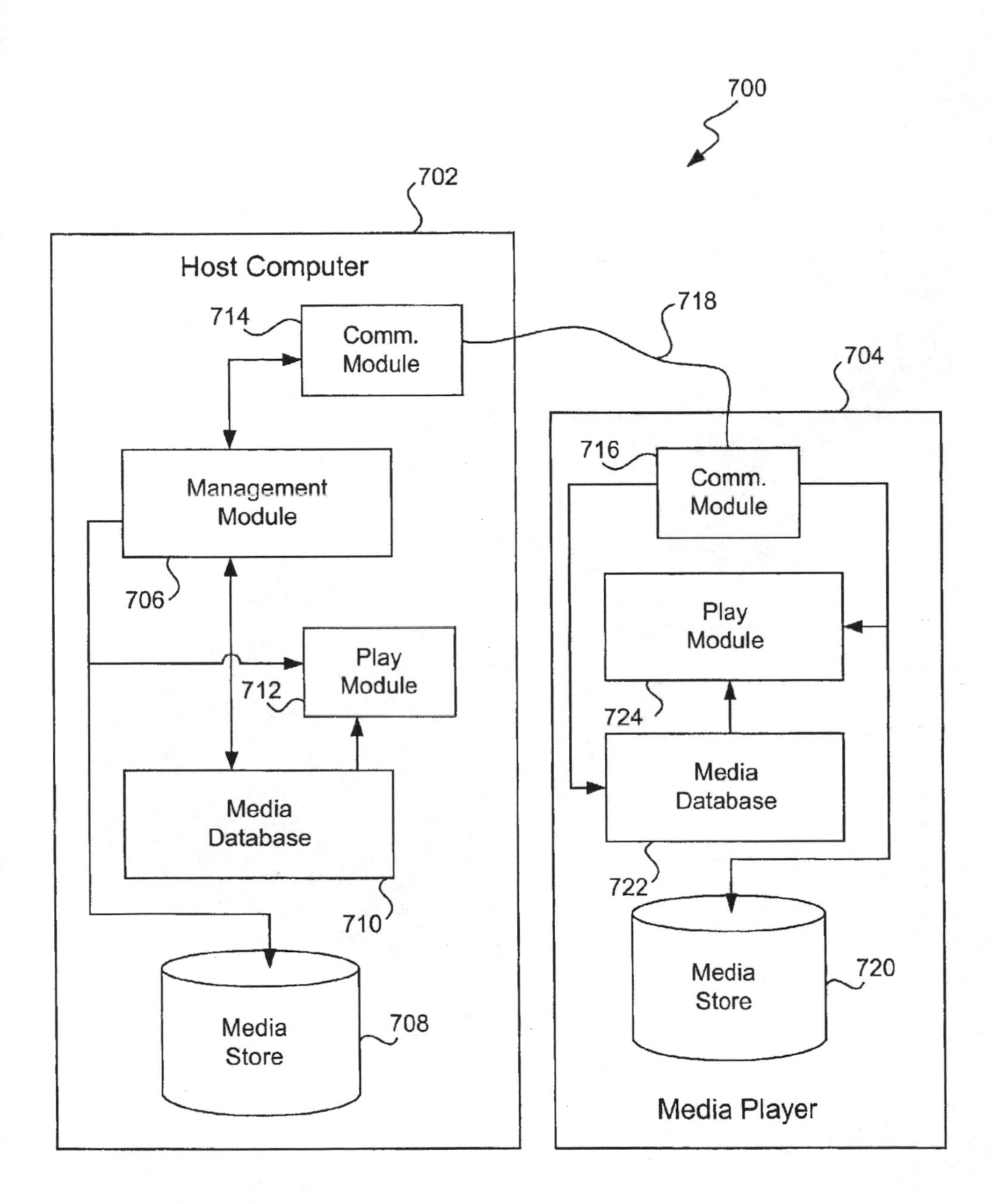

Intelligent interaction between media player and host computer, US 2003/079038, published 24 April 2003

Community Design 000450796-0004,
registered 19 December 2005,
one of many slightly different designs
for the iPod

found it in the form of a British invention where a patent was applied for by Kane Kramer and James Campbell as long ago as 1981, WO 83/01705, the 'Portable Data Processing and Storage System'.

The original invention was not a commercial success, partly owing to what was described by Kramer as an 'attempted coup' to take over his company. The technology was ahead of its time, since the technology of the day allowed only three and a half minutes of music to be stored on a memory chip. The prototype was the size of a credit card, had a rectangular screen and a central menu button to enable scrolling through the contents. Music would be downloaded over telephone lines. Not terribly different, then, from the modern idea of an MP3 player.

Both the European and American patents for the 'Portable Data Processing and Storage System' are now out of protection. While checking the status in the official American database, I was interested to find recorded as a June 2007 'transaction' that the files of the examiner's work and correspondence for this patent were marked as 'lost' and then the next day 'found', so an enquiry had presumably been made about it then.

Kane Kramer tells that he was up a ladder when he was summoned to a phone call from a lady with an American accent. She told him she was from Apple and that the company would like him to give evidence in the Los Angeles court case. At first he thought it was a hoax by friends.

He was flown to America, where he made a deposition and was questioned for ten hours by the Burst.com lawyer. His original notes and drawings had been made in 1979, when he was just 23, and were produced as evidence under the American 'first-to-invent wins' rules. (Patent offices in other countries deem that the first to the patent office wins.)

The case has since been settled out of court. Kramer has been paid an undisclosed consultancy fee.

Fed up with abusive teenagers hanging round your shop?

Many adults dislike seeing groups of teenagers hanging round listlessly at street corners, clearly bored. Shopkeepers in particular dislike the reduced sales that tend to result from youths idling nearby. Howard Stapleton is an inventor who has come up with a controversial solution.

His Mosquito device gives off a high-pitched whine. Because hearing worsens with age, the idea is that only the young will hear it, and so move on to another location. This is a variation on the established idea of endlessly playing Barry Manilow, Mantovani or Mozart, which are all, apparently, guaranteed to drive modern youth to despair.

Stapleton is the managing director of Compound Security Systems, which is based at Merthyr Tydfil. He says the idea originated in an experience he had at the age of 12. He was visiting a Midlands battery factory with his father, who was the chairman of Ever Ready Batteries. The boy could not bear the noise from the ultrasonic welding equipment, which used high-frequency sound to melt and fuse plastics. He had to walk out, but none of the adults had heard a thing.

Stapleton's patent application for a 'High frequency sound teenager dispersal device', GB 2421655, was published in 2006. The specification makes interesting reading, with sentences such as 'The frequency has been chosen to be preferentially unpleasant to younger persons.' Frequencies between 15.5 and 17.5kHz apparently cause a ringing sound in the ears of the young (he experimented on his own children). Sequences of several pulsing tones were found to be most effective. Those aged over 25 are very unlikely to be able to hear them.

The problem itself is familiar to only too many. The first man to try the product was Roger Gough, who runs a food shop in Barry, Wales, with his parents. He had a problem with surly teenagers who used to plant themselves just outside his door. They would smoke, drink, and swear at customers, and make occasional disruptive forays inside. Sometimes they would fight, steal and assault staff. Stapleton offered his device to Gough for a free trial. The loudness of the broadcasting was at set at 75 decibels, which is within government safety limits. The deterrent was the size of a brick and was wall mounted outside the shop, to be used only during certain hours.

At first, the youths repeatedly went into the shop with their fingers in their ears, 'begging me to turn it off', Mr Gough said. But he held firm. 'I told them it was to keep birds away because of the bird flu epidemic.'

One account said it was as if someone 'had used anti-teenager spray around the entrance'. Mr Gough said he was 'very pleased' with the results, and that shoplifting also dropped because 'It's very difficult to shoplift when you have your fingers in your ears.' Gough's children were understandably less enthusiastic.

The device, whose workings are described in the patent application, is called the Mosquito because it's small and annoying. Another trade mark by Stapleton is a sinister looking mosquito.

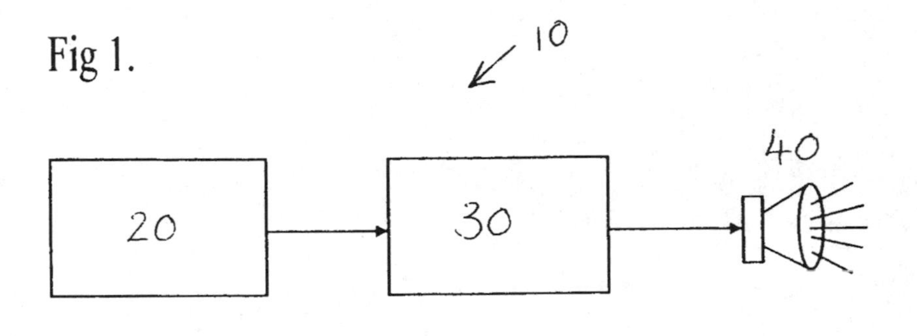

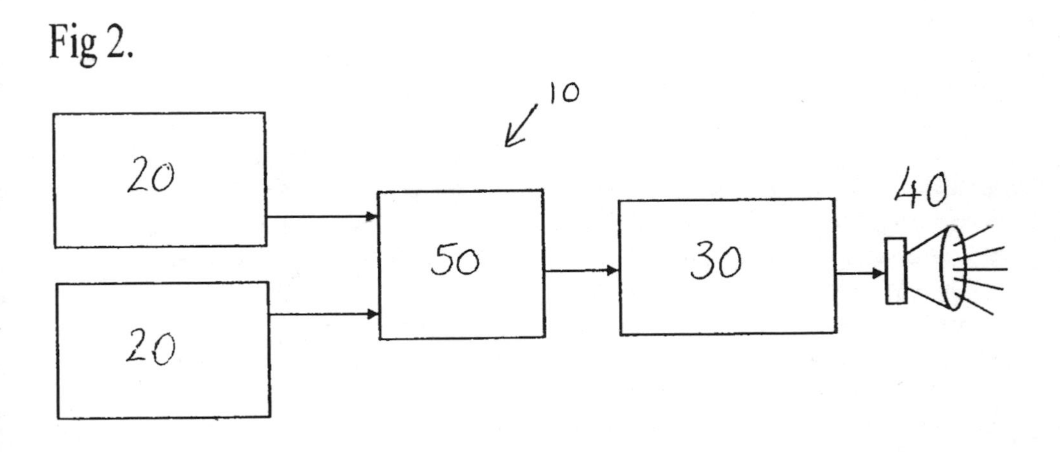

High-frequency sound teenager dispersal device, GB 2421655, 28 June 2006

British trade mark 2418705, owned by Howard Stapleton

The future, perhaps, looks bright for Stapleton, if not for the teenagers. For a while there was a website offering the product, asking the pointed question, 'Fed up with abusive teenagers hanging round your shop?' The official site still warns of 'the eternal problem of unwanted gatherings of youths and teenagers in shopping malls'.

Many police forces and local councils backed the device but the UK Children's Commissioner and the Home Office were not so keen. Nor was the civil liberties group Liberty. 'Buzz off' was the name chosen for the campaign against it, while 'indiscriminate', 'demonising', and 'low-level sonic weapon' were the kind of terms used by those opposing it. In February 2008 the government announced that it would not ban the device, but considered it should be used only 'as a last resort'. So far, over 6,000 have been sold worldwide.

There is an unexpected, and ingenious, coda. Some Welsh teenagers began using the irritating sound as the ring tone on their mobile phones. They can now be alerted to their text messages while the teacher remains oblivious. Use of the ring tone, nicknamed 'Teen Buzz', spread rapidly through British classrooms and then abroad. When Stapleton's 16-year-old daughter Isobel came home with the ring tone on her phone he realised that there was extra money to be made. He quickly came up with the 'official' ring tone, Mosquito Tone, now available via text message for £3, which he claims is superior to other versions. Stapleton's father, still active as a director, replied to his son's suggestion that the ring tone would be the icing on the cake with the riposte that it could well be the cake itself.

Apple's iPhone

On 9 January 2007 Apple's Steve Jobs announced, 'We are going to make some history together today.' He then spent 80 minutes talking about the company's new product, which he said was three revolutionary products in one – an iPod, a phone and an Internet communication device – which worked in a new way. It was to be called the iPhone, and Apple devotees came to realise, in their fervour, that this would be the solution to all their problems.

Two days later a leading IT company, Cisco Systems, said that it would sue Apple for trade mark infringement for using the iPhone name.

Just before that storm broke I had wondered about the word as an American trade mark, and had noticed that Cisco owned it for 'hardware and software for providing integrated telephone communication with computerised global information networks'. They had been using the mark since 1997 and had registered it in 1999.

In fact, claimed Cisco, Apple had attempted to get rights to the iPhone name several times. Cisco's lawsuit claimed that, after Apple's request had been declined, a front company had been established to try to acquire the rights another way. Apparently there had been negotiations up to the day before Jobs announced the product. The name was later registered successfully for certain activities as a US trade mark by Apple, while Cisco retains it for use in the field of contacting 'computerised global information networks'.

The iPhone uses an innovative menu system. As the user moves around the menus, the touch screen alters to show all possible options, rather than confusing the user with invalid ones. Bright icons, 'media mixer controls', show up against a black background.

So how precisely does it work? The idea was plainly to make a combined device that was simple and intuitive to use, rather than requiring a doctorate in computer science, as so many electronic devices do. Jobs has claimed that people dislike using a stylus, as on some devices, and many would agree with that.

Patents are an obvious source of information since they must explain how the invention works, though admittedly for those 'skilled in the art'. However, complex electronic devices incorporating software are usually covered by many new patent applications plus, perhaps, old patents and non-patented concepts, making tracking all the details of how the technology works very difficult. Finding relevant specifications is made more awkward by the use of language that is perhaps best described as 'nerdy'.

This invention is no exception. Persistent Internet searching usually unearths information, or at least suggestions, offered by enthusiasts regarding the patent documents related to a concept. There are four world applications by the company for gesture- and touch-sensitive surfaces or devices. There is also world application WO 2006/020304, 'Mode-based graphical user interfaces for touch sensitive input devices'. This has a flow chart showing that if a touch is detected then the 'user interface mode' is identified. The appropriate GUI (graphical user interface) is displayed and enabled.

Interesting though it is, this is not the key patent specification. The engadget.com site talks of 'a mammoth document that can only be described as the iPhone Patent'. Mammoth it certainly is at 371 pages, 302 of which are drawings, and the site thinks the inclusion of Jobs as an inventor is significant (it is distinctly sceptical about his actual involvement).

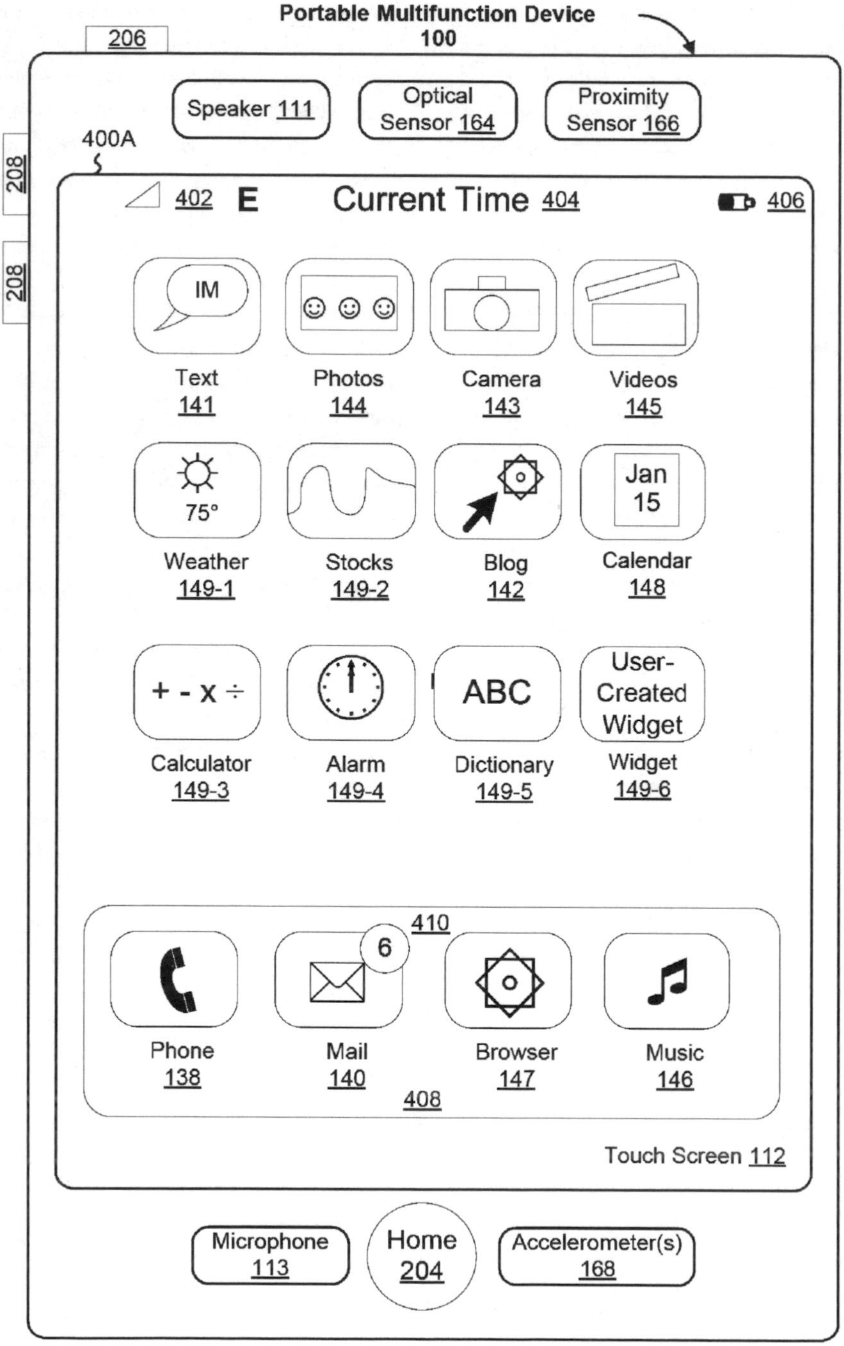

Portable Multifunction Device
100
206
Speaker 111
Optical Sensor 164
Proximity Sensor 166
400A
208
208
402 E Current Time 404 406
IM
Text 141
Photos 144
Camera 143
Videos 145
75°
Weather 149-1
Stocks 149-2
Blog 142
Jan 15
Calendar 148
+ - x ÷
Calculator 149-3
Alarm 149-4
ABC
Dictionary 149-5
User-Created Widget
Widget 149-6
410
Phone 138
6
Mail 140
Browser 147
Music 146
408
Touch Screen 112
Microphone 113
Home 204
Accelerometer(s) 168

US 2008/0122796 is more recent than the previously mentioned application and therefore incorporates later improvements. In fact the equivalent world application was published 10 weeks earlier, so announcing it had 'just been published' was not quite the coup the site thought it was.

Usefully, the application lists numerous related patent applications by the company in the initial page of the description. Many of the drawings show actual or, intriguingly, possible envisaged uses of the device, including 3G functions. Only actual users (I am not one) will appreciate the differences, and some may already have been incorporated in newer models such as the 3.0 (it takes 18 months for a patent application to be published). The drawing displayed here shows an 'exemplary user interface for a menu of applications on a portable multifunction device in accordance with some embodiments'.

The numbers on the drawing refer to references in the text. Unusually, the reader is again and again referred to other specifications for more details on, for example, the proximity sensor (166) and the accelerometer (168). The former is to do with detecting an object and then carrying out an action. For example, if the phone were held against the owner's ear, the screen would go black. The latter was meant to detect a possible theft and trigger an alarm if the phone sensed *very* quick movement. Don't run for the bus then.

Sales of the iPhone have been huge. Over 13 million were sold in 2008, and one prediction is that by 2012 lifetime sales will exceed 80 million. This is because it combines so many functions in an attractive, easy-to-use product. Functionality may be enhanced by over 90,000 specialist 'apps' (applications), software that can be downloaded for a small fee or even for free. Frequently these provide data from the Web to cover a specific interest such as sports results or stock-exchange data, or will perform a simple task such as working out tips on a restaurant bill. Predictions from research firm Gartner are that 4.5 billion apps will be downloaded in 2010, at a cost of $6 billion.

Touch screen device, method, and graphical user interface for determining commands by applying heuristics, US 2008/0122796, published 29 May 2008

The Gocycle bicycle

Cycling around cities has become much more popular in Britain in recent years. Cycling long distances can be tiring, and often there is a need to be able to fold up the bicycle. Neither problem is new, and both foldable and electric-assisted bicycles have been around for a long time. The first effective foldable bicycle is probably the patented design by Harry Bickerton, which dates back to 1974. American patents for electric bicycles go back to the 1890s. The difficulty has always been that the batteries are quickly drained unless used sparingly, for example when going uphill or making a fast getaway at traffic lights.

Richard Thorpe, a London-based design engineer and a former race-car designer for McLaren cars, started his own company, Karbon Kinetics, to commercialise his light-electric vehicles. The first of these was a folding electric bicycle. An early patent from 2004, illustrated here, focused on the machine's foldability.

The wheels were detachable and could be reattached to other points on the frame, so that they 'trailed' and the whole folded bicycle could be wheeled along rather than carried.

Later innovations by Thorpe incorporate shock absorbers, a bicycle seat post, and a motorised single-leg front fork. In 2007 he applied for a patent for an electric bicycle, known as the Gocycle, which claimed to overcome the problems of earlier electric bicycles.

The Gocycle patent application explains that electric bicycles are usually fitted with an electric motor that drives the front wheel, the rear wheel or the pedals. If motors or gearboxes are fitted directly to the front wheel, they are usually mounted within the hub. This means that a puncture necessitates the removal of these parts. The wheel is heavy and unwieldy, and tools are needed to undo the sturdy axle torque controlling nuts. The diameter of the front wheel is much larger than that of a normal bicycle, so that, among other problems, it is difficult or impossible to mount standard-sized disc-brake rotors on the hub. The user also has to retain the heavy motor even if not wishing to use it. It is difficult to remove, and a spare wheel would be needed.

The company claims that its inventions resolve all these problems. The motor's battery is recharged from the mains in about three hours and can sustain the bicycle for up to 20 miles without pedalling. Recharging consumes roughly the amount of power a 100-watt light bulb uses in under two hours. The maximum speed is 15mph, which is the fastest permitted in the UK for an unlicensed electric vehicle. Other innovations include a completely enclosed chain, so there is no danger of oilstained clothing, and a light motor mounted on the front fork assembly rather than the wheel itself, so that the wheel can easily be removed.

The saddle moves further back as it is raised, making more leg room for tall users, and the handlebars can be adjusted for reach as well as height.

The frame is made from super-lightweight injection-moulded magnesium alloy, which helps to keep the weight down to only 16kg. This is much lighter than most electric bicycles, although still heavier than most pedal-powered models. At over £1,000, its price is similarly weighted.

There is a rear shock absorber to ease the ride over potholes. The rear hub has a three-speed gearbox operated by a twist-grip shifter. A handlebar-mounted 'boost' button operates a

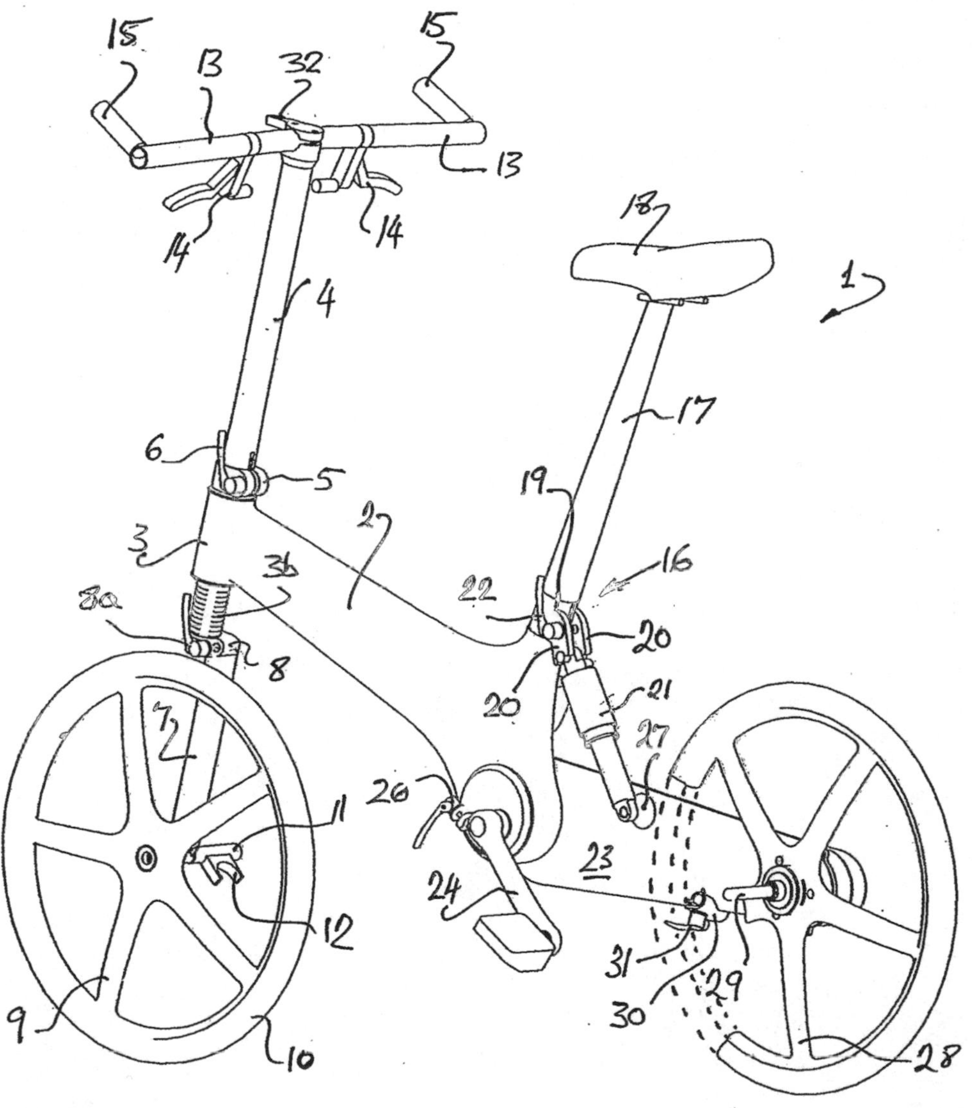

Trailable folding bicycle,
WO 2004/087492, published
14 October 2004

quiet 250-watt hub-mounted motor that drives the front wheel via a tiny ceramic clutch.

The Gocycle, which won the Gold Award at the 2009 Eurobike show, has been described as 'an electric bike that won't make you feel like an utter dork on your daily cycle – the perfect commuting steed for the perspiration-phobe'.

The company has also taken out eight design registrations so, unusually, the three key elements of any invention (patents, designs, trade mark) are all covered effectively.

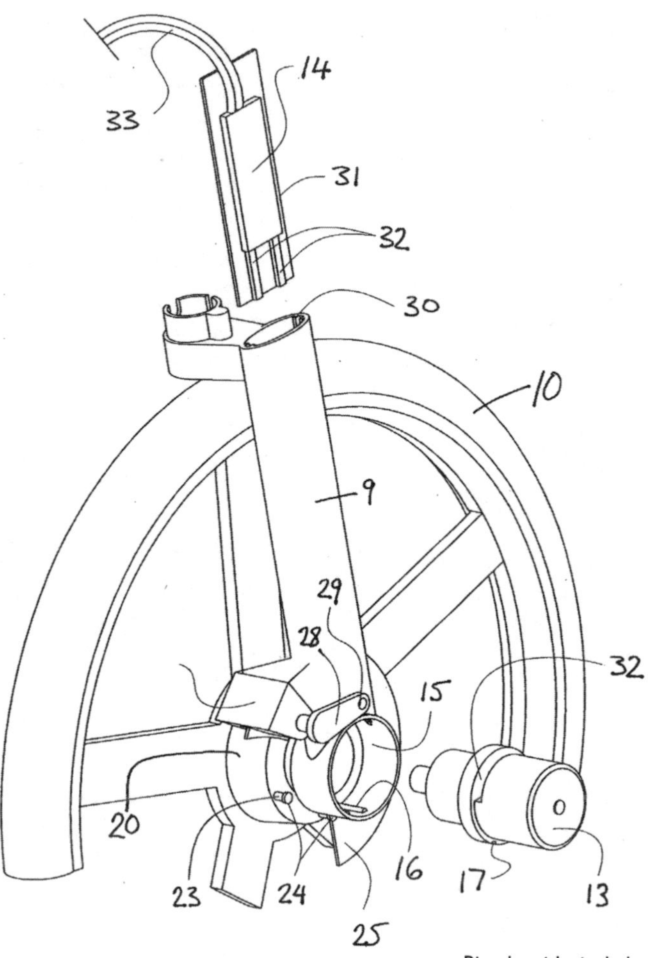

Bicycle with single leg front fork, WO 2009/027683, published 5 March 2009

The Magnamole cable guide

Sharon Wright was a single mum who had just moved into a new house in Scunthorpe. While her telephone and satellite TV services were being installed, she noticed that the engineers had difficulty passing cables from the exterior of the property to the interior. The house had cavity walls and the cables tended to coil into the cavity rather than pass straight through.

'They both fetched wire coat hangers and taped the cable to one end,' says Wright. 'But pushing it through like that was difficult and messy and they left unsightly holes. I also thought it was a health and safety problem. There could have been a mains cable inside the wall and if the drill had nicked the protective sheath, then poking around with a coat hanger would have been dangerous. I decided there must be a better way' – the eternal cry.

A gifted saleswoman, Wright had left a career in estate agency eight years earlier and joined an adhesive-tape manufacturer as UK sales manager. She had just been asked to source a magnetic tape for a client and wondered if magnets would ensure safer and more efficient cable installation.

Cabling has memory and will try to coil back if attempts are made to push it across a gap in a straight line. Something extra was needed to get it through a cavity wall. She devised a wand with a magnetic tip which can be pushed from the inside to the outside through pre-drilled holes in the inner and outer walls. A magnetic cap is then placed over the cable end which is attracted to the wand's magnetic tip. The wand is then pulled back through from inside, dragging the cable with it. There is even a fluorescent tool end, so that it can be used in, for example, an unlighted loft space. It was so simple she could not believe it had not been done before.

She went to a patent attorney, who told her it was a 'no brainer' and perfectly patentable.

For investment, Sharon applied to the *Dragons' Den* television show. By the time she appeared on it she had sold 36,000 units, had an order from BT for 16,500 more, and was making sales in Europe and the USA. Her presentation has been called one of the best ever. She was quietly confident in her product and remembered her statistics. It is rare that all the Dragons are impressed. An objection raised by one was that surely she would saturate the market – a workman needed just one. She replied that she was working on a product for use within cavities that would require a Magnamole to retrieve it. 'That told you,' another Dragon gleefully said.

In the course of the discussion she revealed that the materials cost her about £1 and she was selling it for £5, a healthy and impressive margin. BT had been very interested from the start. She estimated that they alone would save at least £6.5 million annually by using her product.

She was asking the Dragons for £50,000 in return for 15% of the company. The Dragons asked why she wanted their money if she was already so successful. Wright replied that she was working non-stop and could afford to employ only one person for one day a week; she wanted to release her time for what she was good at: sales.

Two of the Dragons, Duncan Bannatyne and James Caan, offered her £80,000 for 25% of the business, as they believed she needed extra finance. She managed to get them to accept 22.5% instead, which in fact meant that they paid more for the equity than her original offer. The Dragons had clearly been charmed by this dragon slayer. 'I was pitching the idea constantly to so many people that I wasn't fazed to appear in front of the Dragons,' she later said.

Wright received 7,000 e-mails after the show was televised in July 2009. These were mostly supportive, but some were orders. A few asked for a date, 'but I have no time for romance'.

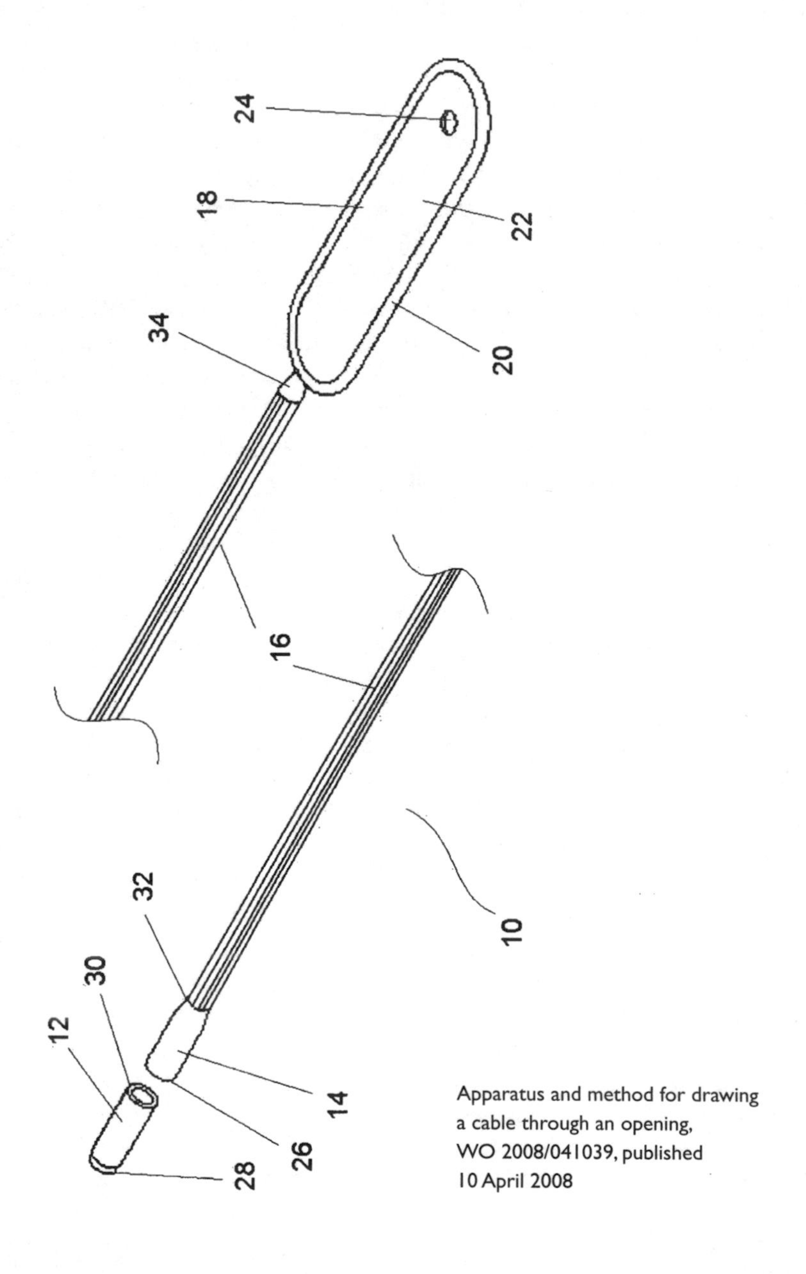

Apparatus and method for drawing a cable through an opening, WO 2008/041039, published 10 April 2008

The Speedo LZR Racer swimsuit

'Faster, Higher, Stronger' is the Olympic motto. One company that has taken the first word, at least, to heart is Nottingham's Speedo International, the well-known swimwear manufacturer.

Their FastSkin II design, an improvement over that used at the 2000 Olympics, was used by the British team at Athens in 2004. This in turn was improved by the Speedo LZR Racer design (illustrated). The company claims that the swimsuit increases speed by reducing passive drag by up to 4% more than the next best suit. That is a lot in competitive sport.

The LZR Racer patent names two women as its inventors: Fiona Fairhurst, based in Nottinghamshire, and Jane Cappaert of Massachusetts. Fairhurst is studying for a PhD at the University of Derby on precisely the subject of this swimsuit. An academic in fluid mechanics acted as an adviser. There was also advice from NASA and sports institutes. Research included analysing the swimming actions of 400 competitors to identify the areas of high friction.

The specification begins by explaining that elasticated fabric has been used to fit snugly against the body. In recent years use has been made of various fabrics with high elastane content which combine different degrees of elastic stretch with a high stretch constant to press more firmly against the body surface for a given degree of stretch. In racing swimsuits this reduces the entry of water between the suit and body, which is a source of drag, and avoids the sliding of the fabric over the skin. It can also reduce muscle vibration, which is believed to be a cause of fatigue and body drag in swimming.

Speedo, though, were 'proposing' in the patent specification 'novel structures for articles of clothing of the kind described enabling improvements in achieving a highly-tensioned fit over the body, especially lower back and abdominal fit, and also preferably taking account of the disposition of muscles over the body'.

The same fabric is used for the whole suit. The trick was to incorporate 'tensioning panel seams' between portions of the fabric to alter the way the suit behaved. Using these reduces the tendency of the fabric to move away from the armpit or groin and means that a closer fit than usual can be achieved in the rest of the suit: the torso and the legs. Streamlining, very important in swimming, is thus improved. The suits also repel water, allow oxygen to flow to the muscles, and hold the body in a better hydrodynamic position. Ultrasonic welding in the seams is used to reduce drag still further. There are drawbacks: the suits are very difficult to get into, the process taking more than half an hour, and are liable to tear.

The new suit launched in February 2008 and had a tremendous impact on times in international swimming meetings. During the following four months 41 world records were broken – 37 of them by swimmers wearing the new swimsuit. Half a dozen or so new records would have been the norm. Many of the record breakers had not previously been highly rated. As of December 2009 only two pre-2008 50-metre pool world records still stood, one each for men and for women.

At the 2008 Beijing Olympics 94% of the races were won by competitors wearing the new suits. In the 2009 World Championships, dubbed the

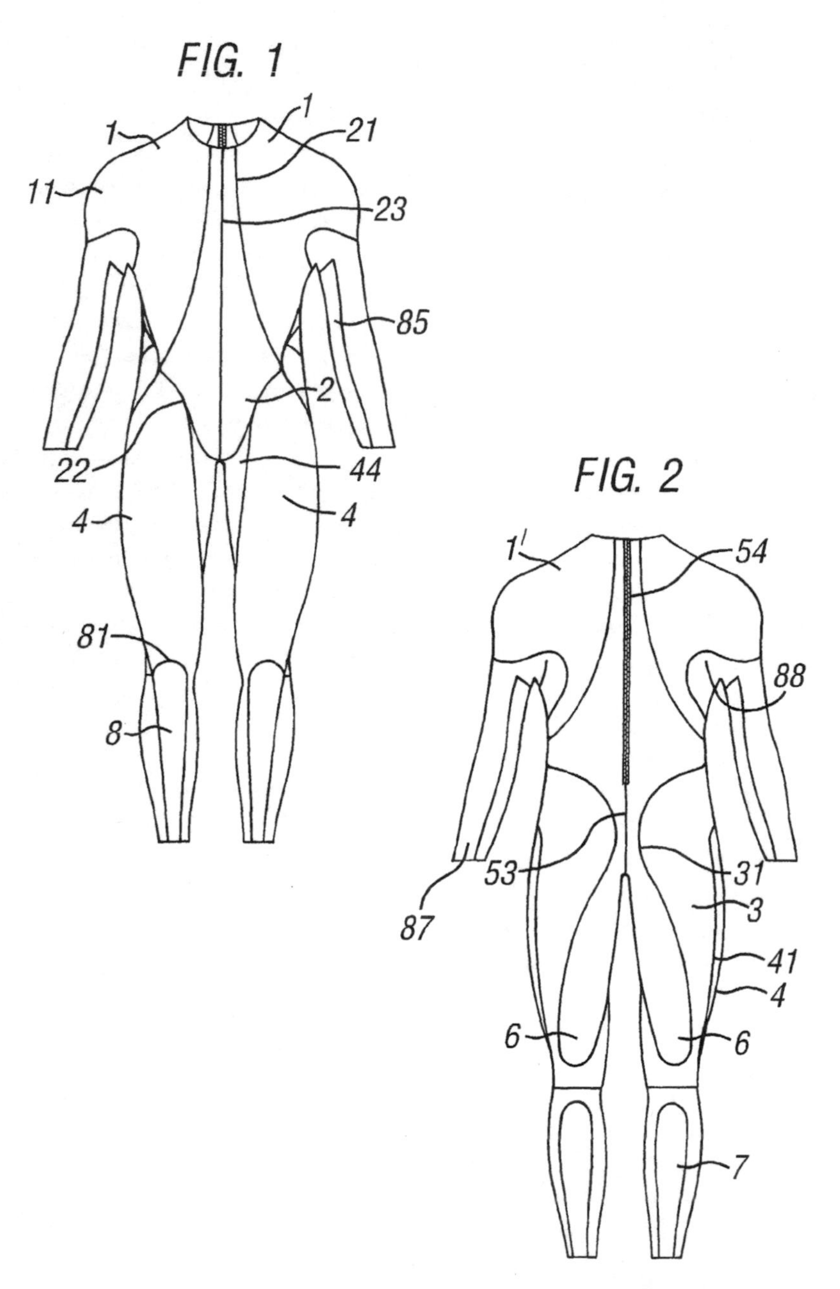

Close-fitting article of clothing with highly tensioned fit, EP 1250858, published 23 October 2002

'Plastic Games', 43 world records were broken, nearly all by competitors wearing the LZR.

Unsurprisingly, all this caused a sensation in the swimming world, and spawned allegations of cheating and 'technological doping'. The International Swimming Federation (FINA) arranged a meeting with Speedo to discuss the allegations and declared them unproven. Many swimmers who had sponsorship deals with other suit manufacturers were often allowed to use the Speedo suits as otherwise they would be at a hopeless disadvantage.

In March 2009 FINA had changed their mind and decreed that from 2010 all such suits would be banned. Non-textile fabrics are now illegal and women's suits have to end at the shoulder straps and above the knee, while men must return to wearing trunks. The new records will stand as valid.

Analogies can be drawn with other sports. New running surfaces such as the Tartan track were permitted as the improved surface benefited all competitors equally, but the 1930s saw a protracted struggle for the introduction of starting blocks, as not all sprinters used them.

A different situation arose in cycling, the outcome of which was very different, too. At the Beijing Olympics, the British cycling team's innovative machinery and high-performance bodysuits resulted in an impressive haul of medals. After the games the cycles were dismantled and the suits shredded to preserve their secrets. However, cycling's governing body has now decreed that in future it will be mandatory for teams to share their innovations. This is, I believe, unprecedented in sport.

The Searaser wave-powered pump

There are numerous inventions in the field of renewable energy. Here is one of them.

Alvin Smith, a retired garage proprietor and lifelong hobby engineer, of Dartmouth, Devon, was in a swimming pool, making little waves, and it struck him how much power was expended in the displacement of the water. *You must be able to harness some of that to produce hydroelectricity*, he thought.

Ten years passed and then he built a prototype in his garden. It was made from four bath wastewater pipes, four snooker balls, a dustbin half-full of water, a steel rod and a plastic pipe. He tried it out in the River Dart. Water was immediately pumped out.

This was the Searaser, a simple arrangement of ballast and floats connected by a piston. As a wave passes the device, the float is lifted, raising the piston and pressurising water. The float sinks back down onto a second float as the wave subsides, pressurising water again on the downstroke.

Most wave-energy devices are sealed and lubricated and use complicated electronics. Searaser is lubricated by the seawater itself and has no electronic components. It is even self-cleaning. Smith describes it as 'Third-World mechanics', something easy to maintain.

'The beauty of it is that we're only making a pump,' he explains. 'All the other technology needed to generate the electricity already exists.'

Searaser either pumps water through a sea-level turbine to generate electricity, or brings it ashore, up to a clifftop reservoir. The pressure is enough to raise the water to a height of 100 metres. It can then be allowed to flow back down to the sea through turbines to generate electricity when needed. This means that power is not available only at certain times but constantly, solving the problem that all renewable energy sources face: variable supply.

The reservoirs would need to be as watertight as possible to avoid saltwater leaching into the soil. Smith envisages double-lining the reservoirs and including an outlet pipe between the two linings. Any puncture in the inner lining would be revealed by water coming from the outlet pipe. A bigger problem might be securing planning permission, even though the reservoirs would not be easily visible, and not visible from the shore. The reservoirs could be used for water sports, shellfish farming, and as a wildlife habitat.

Early trials of a prototype, backed by three local businessmen, were completed in April 2009 off the South Devon coast. Results were encouraging. The prototype used a piston less than a metre in diameter, only a tenth of the intended size, but was able to pump an average of 112,000 litres of water a day for a month, sometimes using waves only 15cm high.

The full-size machine would have a maximum output of 1 megawatt of electricity, enough to supply the needs of 1,700 homes. If the water is stored in the off-peak hours smaller wave heights would be needed, as less output would be needed to supplement the stored water. At average wave swells the output would be approximately half the maximum, utilising pumped-on-demand storage.

Trials of two medium-sized devices off the southwest coast had been planned for early 2010, but these were cancelled due to a lack of funding.

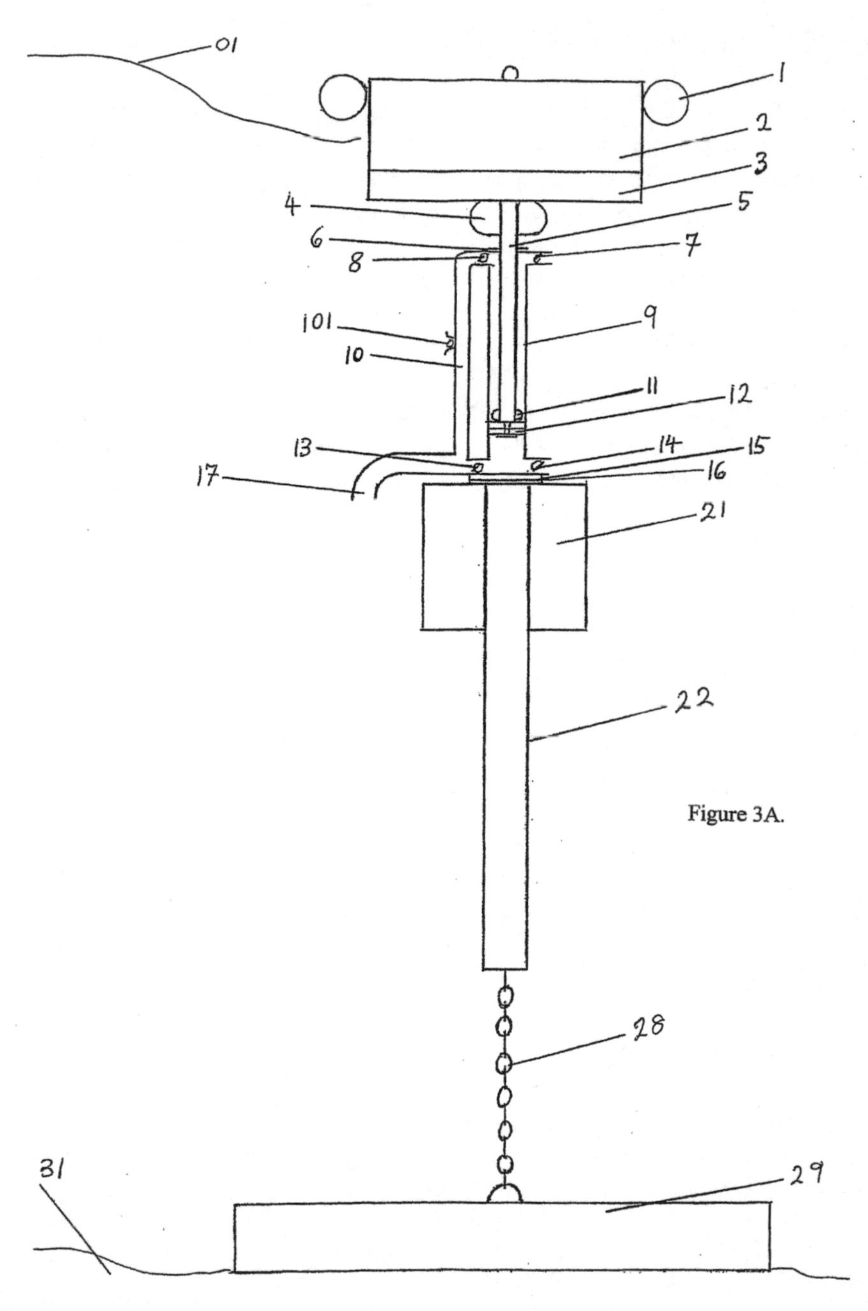

Figure 3A.

Hydro column, WO 2008/090302, published 31 July 2008

Smith is thinking of using the pressurised saltwater to produce drinking water by using reverse osmosis, forcing water through filters that would trap the larger salt molecules. Conventional desalinisation plants use a great deal of power. With Smith's method, 'All you'd have to do is reduce the size of the piston and increase the size of the floats to increase the pressure,' he explains.

A company has been formed, Dartmouth Wave Energy Limited, to exploit his invention. They claim that 11,000 Searaser units would power the entire domestic demand of the UK.

Smith is certainly hoping to see a return on his investment but is also keen to see the technology deployed worldwide as it is so simple to install and maintain and would be very useful in arid coastal regions.

According to the company's design team, costs will be lower than most renewable technologies. The Searaser is the cheapest part of the plant at about £350,000 per unit. An additional £950,000 per megawatt would be needed for piping and the hydro turbine. One-off costs for reservoirs would be additional, but might be ameliorated by the use of existing sites such as disused quarries. This possibility is being investigated by the local borough council on Portland, Dorset and the project has attracted interest from around the world.

The Road Refresher no-spill dog bowl

Natalie Ellis of Leigh-on-Sea in Essex was inspired to create her travel bowl by her pet miniature pinscher, Shizza. On hot summer days, she used to carry the animal on her lap and give her water by hand at traffic lights. But when the lights changed, still-thirsty Shizza would jump around the car, demanding more. The police were unimpressed, and threatened to arrest her for dangerous driving.

To avoid prosecution, Ellis spent months cutting up plastic and foam in the kitchen, trying to design a bowl that Shizza could drink from while the car was in motion. Eventually, she succeeded.

Open bowls easily spill their contents. This bowl is different. A dish-shaped section (24 in the second drawing) floats on the surface of the water, trapping it. Slits in the dish allow some water to well up so that the pet can drink. Because only a small amount of water is on the surface at any one time, there is much less risk of spillage. The dish is unable to rise above an in-turned rim (28), so there are no spills if the car brakes sharply or the bowl is knocked. If the dog is left in the car the filled bowl can stay with it.

The rim itself is snap-fitted so that the dish-shaped section can be removed for cleaning. This in turn fits into the base section. Patent protection has been granted in both Britain and the USA. The trade mark is registered in both Europe and the USA.

Natalie left school at 15 with no qualifications. A decade ago she began working for herself and, after leaving a career in sales, she went into the pet market. She was working from 5 a.m. to midnight daily selling product ranges to supermarkets while raising her teenage daughter Leah. She devised the Road Refresher and applied for a patent in 2004.

Sales were good enough by 2008 for her to seek additional capital to finance increased production of the bowl for the huge American

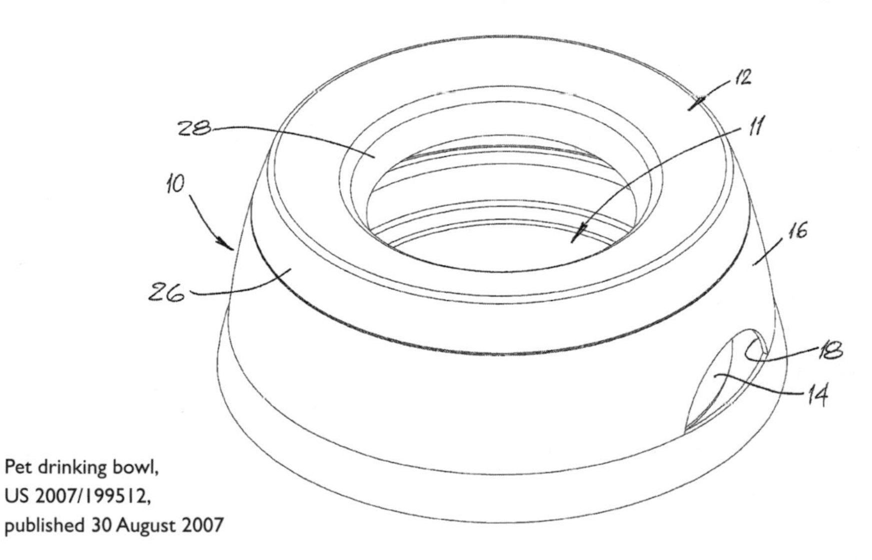

Pet drinking bowl, US 2007/199512, published 30 August 2007

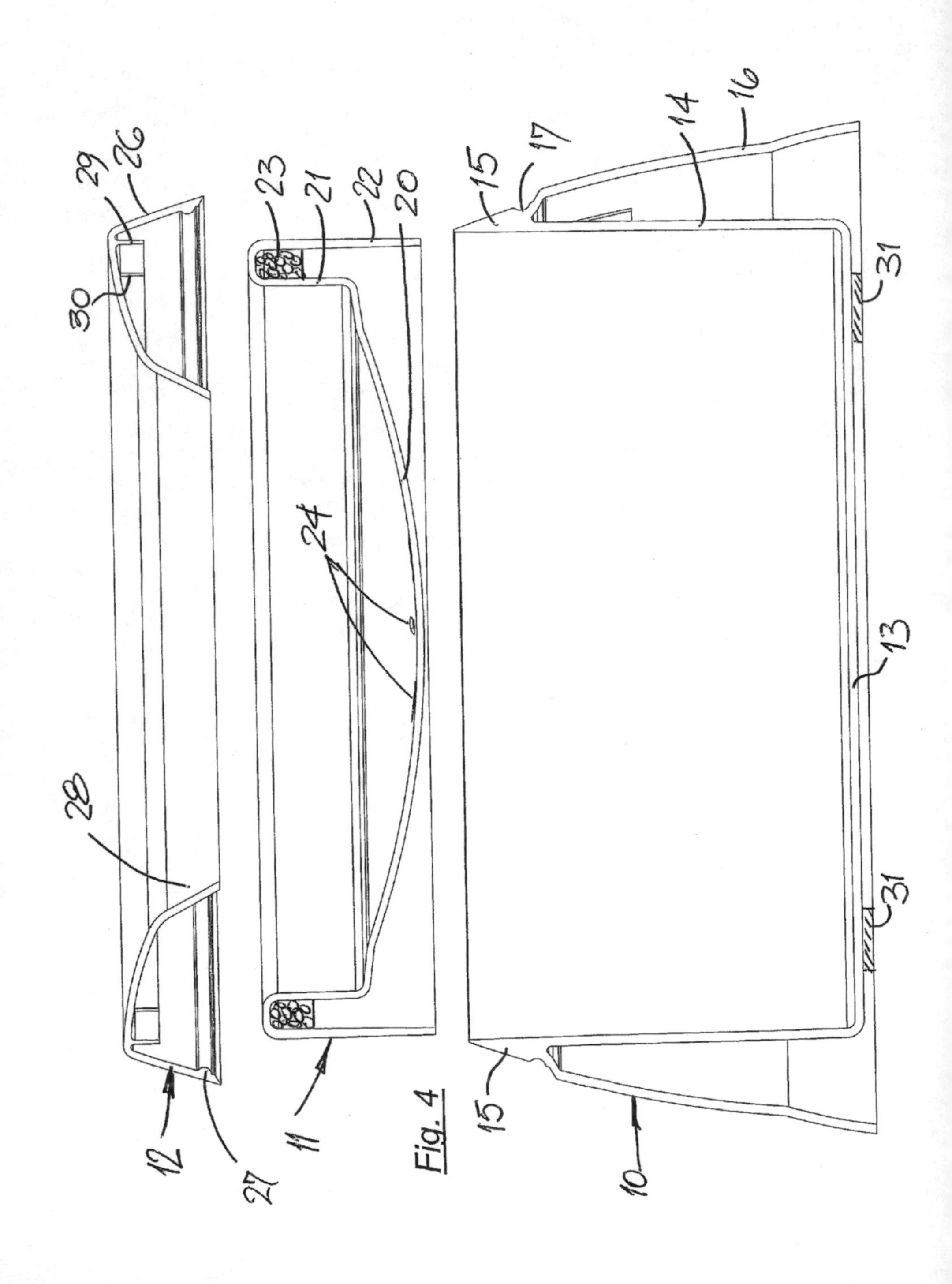

Fig. 4

market. She appeared on *Dragons' Den* to ask the five Dragons for £120,000 in return for 15% of the equity.

The Dragons certainly liked the product. Figures were confidently presented: sales had been 36,000 in the last five months, with a turnover of £75,000 and gross profits of £42,000 and £13,000 net profit. The money was needed for new moulds for an extended range and to tool up so that her company, Prestige Pet Products UK, could make more than the current 8,500 units per month.

When asked about her track record, Ellis explained that she had had to wind up a company she had previously run. There had been a discrepancy in her VAT return when she had entered a VAT invoice for £17,000 twice, though she maintained that she was not entirely to blame. One of the Dragons, Theo Paphitis, pressed for more details. And the story of why the company had failed tumbled out.

She had suffered a transient ischaemic attack – a mini stroke.

She was clearly distressed as she recalled the event, and turned away and said just enough to shock and move the Dragons. She has later told how the stroke occurred at a trade fair in Germany in May 2004. She had suffered from amnesia for a couple of months, lost feeling on her left side, could not walk, and was still affected many months later. She could no longer use her computer because she had forgotten her password, would lose concentration while talking, and would forget where she was while driving. She slept a lot and her body more or less shut down. Natalie later said, 'I didn't want to tell them [the Dragons]. It's a part of my life I have put in a box. But I was being asked again and again and I had to tell the truth. It was like reliving a nightmare and I burst out crying.'

It had been 18 months before she was able to work again at a slower pace. The 19-hour days selling pet accessories to major supermarkets were over, as the contracts had been lost.

More gently, and showing their admiration – they said it must have been like 'climbing Everest', and that she appeared to have recovered completely from the stroke – the Dragons began to explain why they did not want to invest. It was a 'lifestyle' product with a small market, and not an investment opportunity. America was the graveyard of British business. She needed infrastructure and not finance. She should team up with a large company. So she departed without the requested finance.

As soon as she got home she began to bombard American pet stores with e-mails, and set up a website to advertise the product. Nine months later, business was booming for the Road Refresher. She had sold 100,000 bowls, a million pounds' worth, to America in only four months. Then White House officials, having heard a talk-show host recommend the product, asked if President Obama's dog Bo could be provided with one. A special version, modelled on the White House, with an American flag and Bo's name in glittering diamanté, was devised.

The future looks bright, and Ellis says, 'When I go back to *Dragons' Den* I will be sitting in one of the chairs.' Shizza passed away from old age in 2006 and is very much missed, but she has left an enduring legacy.

Nintendo's Wii game console

Nintendo's Wii game console has been a huge hit since its launch in 2006. As of September 2009 over 56 million units had been sold worldwide. Demand remains high: the console sells out almost immediately in the UK as soon as supplies are delivered to shops, despite its price tag of around £300.

The provisional name was Revolution, but it became Wii after the company decided that the new name was short, to the point, easy to pronounce and distinctive. It means the idea of 'we', us, or, it has been suggested, two people side by side in the double letter i. The games are often designed for more than one to play. Certainly, the name has provoked some lavatorial humour in the UK.

For the benefit of those who have somehow missed the excitement, the console works on a new principle. Formerly, a console was designed so that pushing buttons, pulling on a joystick and so on controlled what happened on the screen.

Wii differs in that the console itself is waved around. In a game of tennis, for example, the player mimes making a stroke and the on-screen character hits the ball. Several people can play at the same time, or a single player may compete against a game-generated character. Its use as an aid to fitness has been promoted too, extending its demographic beyond the usual youthful game-player profile. I can testify that it is very addictive.

A remote-control console with buttons on it, in shape rather like a bigger version of a typical TV remote, was originally the only way to play. Following damage to consoles caused by excited players gripping too tightly or hitting them against hard surfaces, a special glove was introduced for many games.

While researching this book, I often made use of Internet fan sites that suggest relevant patents for electronic products, but in this case I drew a blank. I did discover, however, that Nintendo has been threatened with court action for infringement of American patents.

One company that pursed this avenue was was Interlink Electronics, who alleged infringement of their US 6850221, 'Trigger operated electronic device'. Sources on the Web said that it was indeed rather similar to the Wii principle. One of the things that you can do on the esp@cenet patent database is to ask, when looking at a record, to 'View list of citing documents'. This refers to more recent patent specifications that a patent office examiner cited as similar to the older, known patent document. I duly asked to view, and only one relevant patent document was listed: EP 1757343, 'Game operating device'.

It was by Nintendo, and had been published two years after US 6850221, in February 2007. The console as illustrated does look rather short in comparison with the real thing, but otherwise the drawings look very similar to the product. The document explains that there is an 'imaging information arithmetic unit' at the front end of the console which processes infrared signals containing data about the 'position and/or attitude of the controller' (and its speed) which are sent to LED modules within the sensor bar. The sensor bar is a separate unit which is centred either above or below the television on which the game is being displayed. The sensor bar also calculates how far away the signal is coming from, to mimic forward or backward movement.

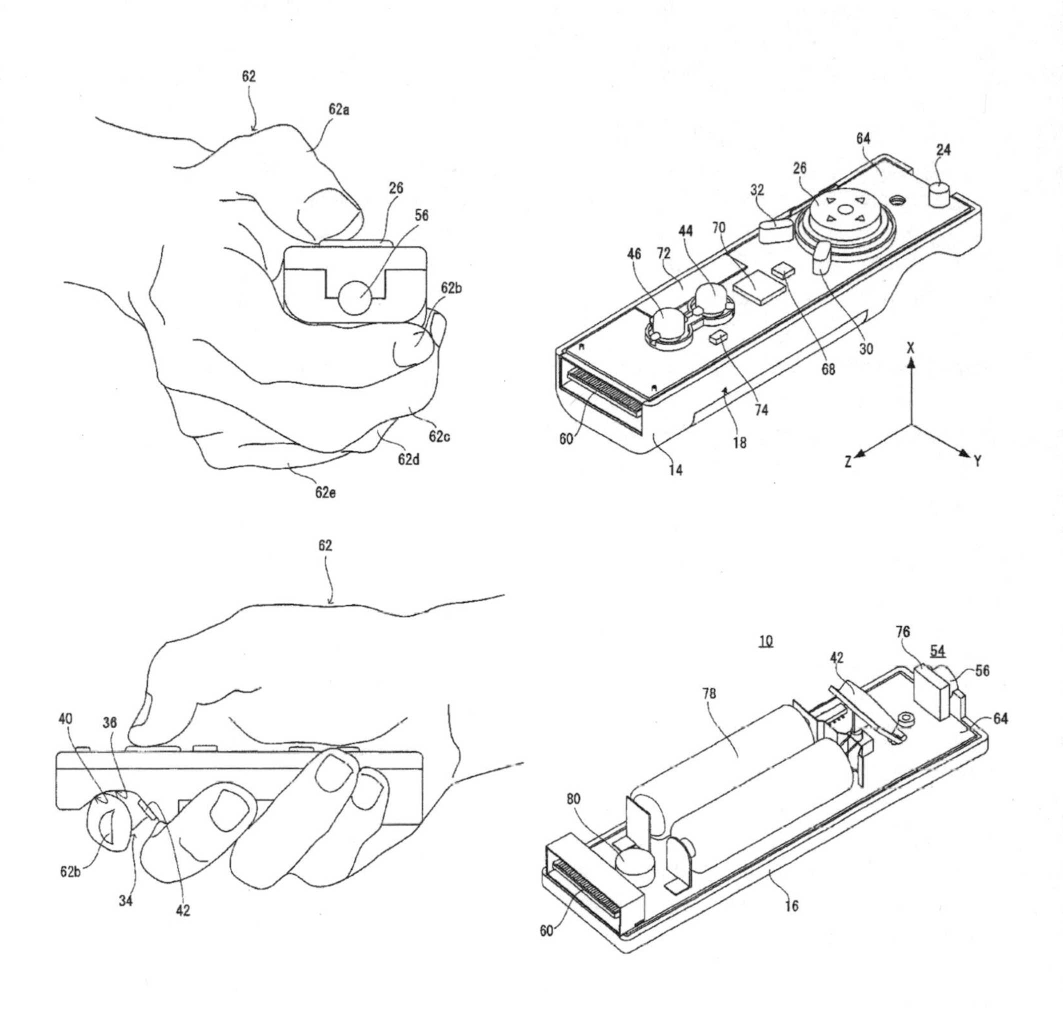

Game operating device, EP 1757343, published 28 February 2007

Nintendo's EP 1832321–322 explain about 'Motion determining apparatus and storage medium having motion determining program stored thereon' and are probably relevant.

Despite the citation by the examiner looking for similar inventions so as to invalidate the new application, confident sources on the Web state that the two patent documents don't have a huge amount in common other than that they are both remotes with buttons that use infrared rays, which is what most do. That could explain why Interlink has decided not to pursue the action.

Wide-image scanning of the retina

Five-year-old Leif Anderson was unable to tolerate drops in his eyes to dilate his pupils for an eye scan. Consequently, a detached retina was not detected and he lost the sight in one eye.

This tragedy drove his father Douglas, a Dunfermline design engineer, to look for a device that did not need drops to make an early diagnosis of eye problems. Conventional examination techniques as carried out in eye clinics only allowed the practitioner to view a limited area of the retina (at the back of the eyeball). Problems around the periphery often went undetected. Available equipment could provide a greater viewing angle, but only with the use of eye drops to dilate the pupils. Anderson thought of using an ellipsoidal mirror to reflect a laser beam and provide a high-resolution digital image of over 80% of the retina in a single scan.

It was fifteen years before a patent was applied for by the company Anderson formed in 1992, Optos. The application is in the name of the inventors who had carried out the detailed development work.

An explanation of how the ophthalmoscope works involves some difficult science. The application itself, referring to the drawing, states as a summary:

> A scanning ophthalmoscope (10) and method is provided for scanning the retina of an eye comprising a source of collimated light (12), a first scanning element (14), a second scanning element (16), scan compensation means (18), wherein the source of collimated light, the first and second scanning elements and the scan compensation means combine to provide a two-dimensional collimated light scan from an apparent point source, and the scanning ophthalmoscope further comprises scan transfer means (20), wherein the scan transfer means has two foci and the apparent point source is provided at a first focus of the scan transfer means and an eye (22) is accommodated at a second focus of the scan transfer means.

Optos's Panoramic200 model captures a high-resolution image of 80% of the retina in a quarter of a second. Previous methods typically captured much less information from a small part of the retina.

Local venture capitalists provided finance for the early stages of development. More money was needed for large-scale production and international marketing, and so in 2006 the company was floated on the London Stock Exchange and raised £50 million, with Anderson as a director.

Sales in the year ending September 2004 were £16 million, and this had grown by 2008 to £55 million. It has sold particularly well in the large American market, where there is a ready acceptance of new diagnostic technologies and of screening generally. As well as its use in detecting eye disease, the Optos scanner can detect a wide range of other conditions, including diabetes, high blood pressure and even some cancers. Optos hopes to encourage the medical profession to adopt the system to provide routine check-ups.

The device has now performed over 24 million eye examinations. One of these saved the sight of Leif's other eye. Not surprisingly, Leif now works as an engineer in a Scottish medical electronics company.

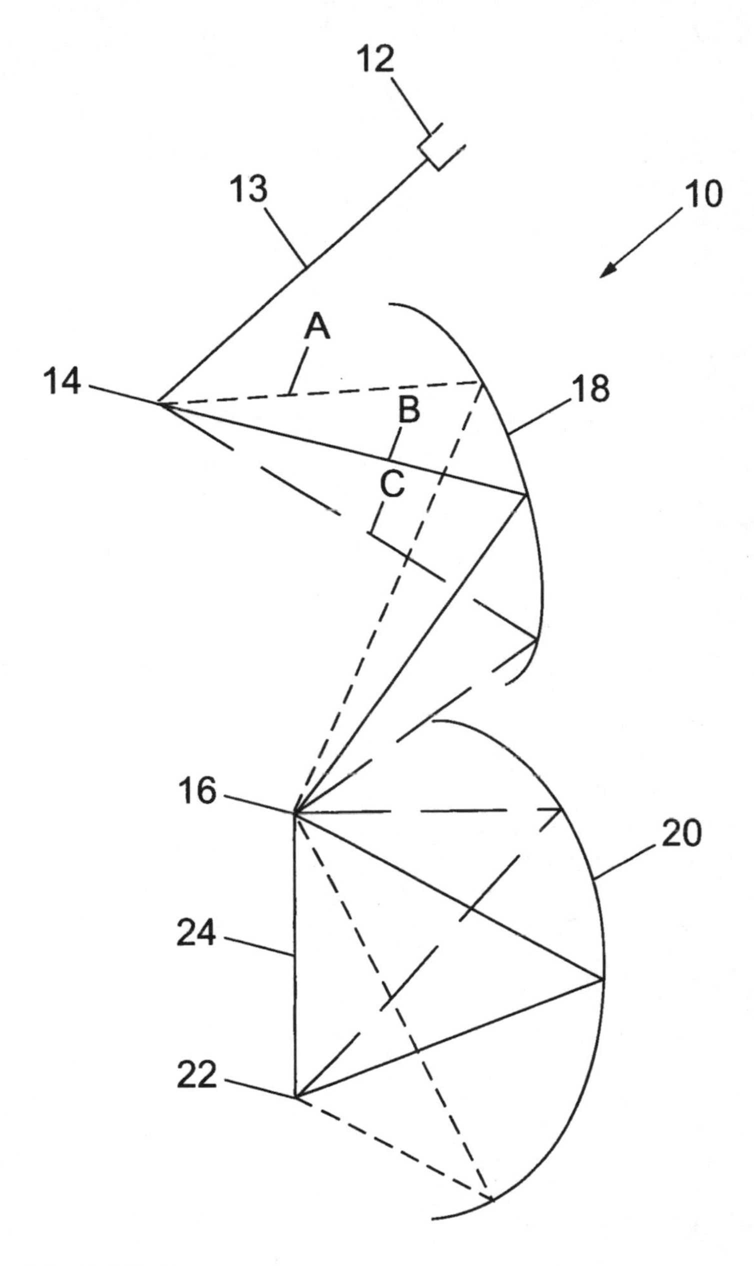

Improvements in or relating to scanning ophthalmoscope, WO 2008/009877, published 24 January 2008

Who invented the MRI? A Nobel Prize dispute

It was announced on 6 October 2003 that that year's Nobel Prize for Medicine had been jointly won by Paul Lauterbur, a professor at the University of Illinois, and Sir Peter Mansfield, a physicist at Nottingham University.

Together these men made possible magnetic resonance imaging (MRI) techniques as used today. MRI was an important advance over body scanners, which use X-rays, because the latter cannot easily recognise soft tissue. Originally the name was preceded by 'nuclear', but this was later omitted as it frightened patients. Lauterbur was the first to produce images showing different types of water (such as heavy water, which contains an isotope of hydrogen with one extra neutron particle). He has four patents, but commented that Stony Brook University decided it was not worth patenting his technique. 'That turned out not to be a spectacularly good decision,' Lauterbur said in 2003. It took him a decade to get the US government to finance an early MRI machine.

Mansfield has 14 patents filed from 1990 onwards for the University of Nottingham, which have reportedly brought him valuable royalties. He has published over 50 papers in the field since 1965. The first advance he made was to show that it is possible to produce a selective image of a particular two-dimensional slice of an object held in a graded magnetic field by delivering a blast of radio waves at a designated frequency. He also developed techniques to manipulate the applied magnetic fields, and algorithms to interpret the resulting mixture of radio signals so that images could appear in a few seconds. His work has allowed researchers and clinicians to take snapshots of fast-moving events, such as the beating of the heart, or to scan blood flow through the brain.

A Nobel Prize for MRI had long been anticipated but, because its development involved many steps, deciding who should receive the prize must have been difficult. 'It's been a hot potato for the Nobel Prize committee,' suggested John Griffiths of St George's Hospital Medical School, London, who uses MRI to diagnose and study cancer, one of its major clinical applications.

The 'hot potato' indeed came about due to the reaction of Raymond Damadian. In 1972 he had applied for what became US 3789832 with an attractive drawing showing a patient standing within a vertical structure. This is normally credited as being the first patent for the MRI concept. It was improved by his US 4354499 with the now-familiar tunnel scanner. He immediately found out about the award by being on the Nobel Foundation website for the 5.30 a.m. announcement of the 2003 prizewinners. (Lauterbur, incidentally, had been informed just two hours earlier, when he was woken by a phone call from Sweden.) Damadian's response was, three days later, to place full-page advertisements in the *Washington Post*, the *New York Times* and the *Los Angeles Times*, headed 'The Shameful Wrong That Must Be Righted', and quoting scientists in his support.

Damadian's advertisements asked readers to cut out a coupon and to send it to the Nobel committee demanding that he be recognised as the third joint winner of the prize. The advertisements were paid for by his MRI imaging company, Fonar. It has been estimated that the *Washington Post* spread alone cost $80,000.

Coverage of his fury continued to be available, years later, at the Fonar company website, listed as 'Expressions of Outrage Over the Nobel Prize in Medicine for the MRI by The Friends of Raymond Damadian'. The awards were 'an affront to our Presidents Ronald Reagan and George H. Bush', under whose administrations Dr Damadian had been awarded the National Medal of Technology and been inducted into the Patent and Trademark Office's National Inventors Hall of Fame. Patent attorneys were said to have stated that the decision to 'ignore' Damadian was an 'affront' to the American patent system as his patent had been affirmed by the Supreme Court.

A favourable response, or indeed any response, never occurred, as Nobel committees refuse ever to discuss the justification of their awards. Nor did Lauterbur (who died in 2007) or Mansfield comment. It seems likely that Damadian was motivated more by professional pride than by the $1.3 million prize money that Lauterbur and Mansfield shared, for, he has said, 'I know the outcome of this is to be written out of history altogether.'

Leaves on the line

Every autumn the British public make fun of the famous 'leaves on the line' explanation for delayed trains, and complain that only Britain has this problem. In fact railways across the globe face similar challenges, and it costs some £100 million annually in Britain alone to maintain the tracks in good order.

The problem is that wet and decaying leaves leave a slippery residue on the track, and that makes high speeds dangerous. They can also result in the presence of a train not being detected by track-based signalling systems. Water-jetting methods currently used to scrub the tracks clean are slow, unreliable and expensive. Formerly brake shoes would grip the wheels and therefore clean any residue from them (rather than the track) when the brakes were applied, but modern railway brakes use a different design.

Malcolm Higgins, a former Royal Navy officer from Lee-on-the-Solent, Hampshire, was driving down a motorway one day in 1999 while listening to a radio discussion of this problem. 'I am no scientist or railway expert,' said Higgins, 'but it suddenly occurred to me that lasers, which are so common nowadays, ought to be able to help.' His concept was to attach high-powered lasers to trains which would clean the tracks as they passed over them.

Within months he formed a company, LaserThor, and patent protection was granted in Britain and the USA.

Higgins also paid for researchers at the Defence Evaluation and Research Agency (DERA) laboratories at Great Malvern to see whether rails could be cleaned by lasers without causing damage. The answer was that it could be done in principle, but that it wasn't practical for a railway system. A laser cleaning unit would take up an entire carriage and would need an impractically large power supply.

Higgins then persuaded Railtrack (now Network Rail) to look into the idea. They are the company responsible for Britain's rail tracks, and some of their staff had witnessed the DERA trials. They helped LaserThor to obtain the necessary safety clearance to carry out trials in the autumn of 2000. This was shortly after a damaged rail had caused a fatal rail crash at Hatfield which had forced Railtrack to check numerous lines for similar defects, paralysing Britain's rail network in the process. 'The fact that we were allowed to run trials even when Railtrack was under so much pressure to deal with the short-term problems that arose in the wake of the [Hatfield] crash shows how enthusiastic the company was,' said Higgins.

Railtrack bought two prototype machines for £1.2 million each on condition that they would be granted exclusive use of the maintenance trains in the UK, and the trials were further financed by a private investor and a government SMART award (Small Firms Merit Award for Research and Technology) worth £40,000. An electrical engineer and the Rutherford Appleton Laboratory in Didcot provided expertise. Spectron Laser Systems, a Rugby-based company, lent LaserThor a pulsed neodymium yttrium aluminium garnet laser. The laser was installed, carefully padded, in a carriage. Mirrors were used to deflect the beam down a pipe onto the track. Power was provided by a 64kW diesel generator in another carriage. The pulses produced a series of tiny explosions that blew the leaf residue off the rails, resulting in a clean track and improved traction. Grease, water and ice were also removed. The track was undamaged at the 8 kilometre per hour speed that was attained.

A faster speed was, of course, desirable. Higgins contacted the Fraunhofer Institute for

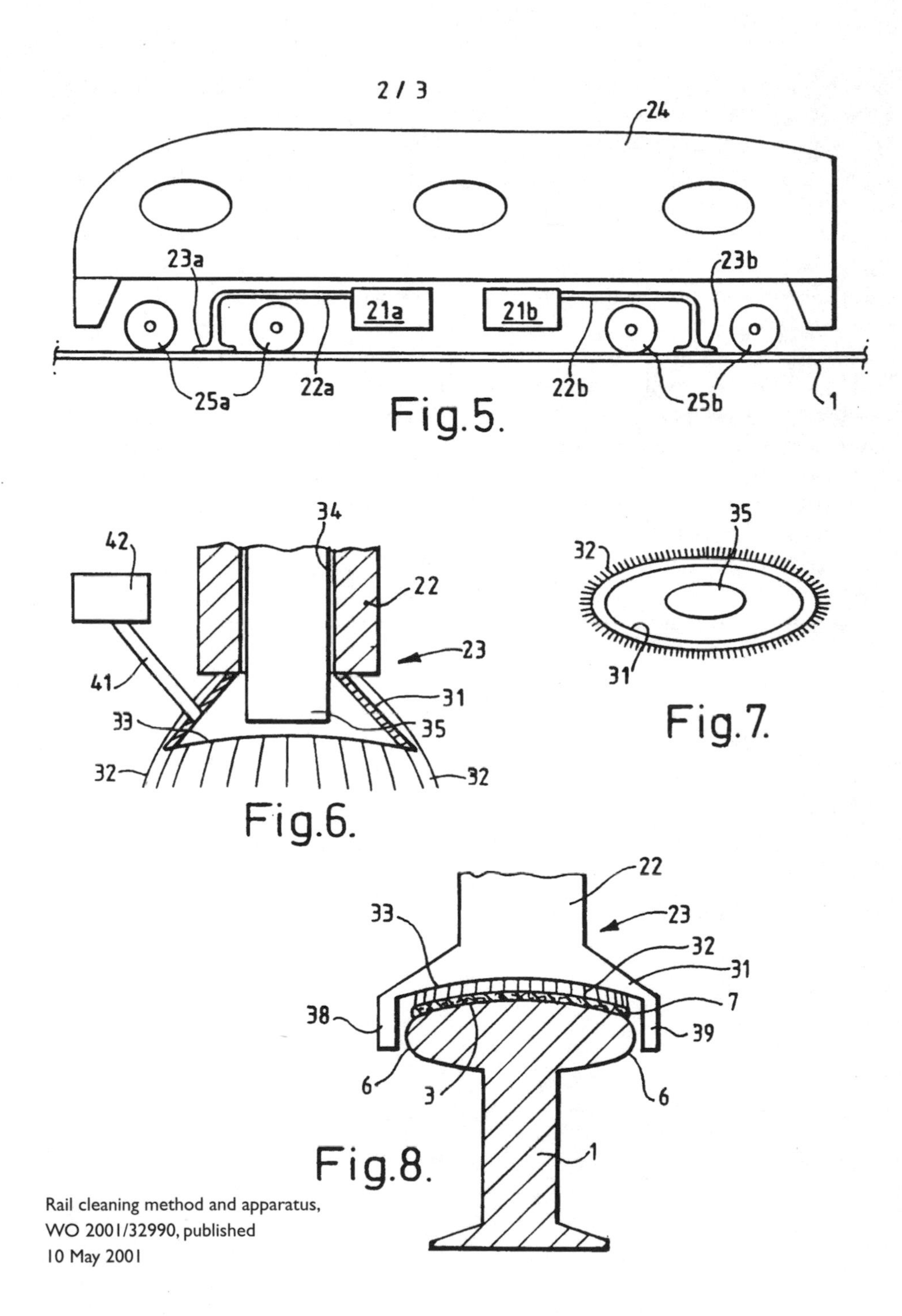

Rail cleaning method and apparatus, WO 2001/32990, published 10 May 2001

Laser Technology in Aachen, Germany, who built a laser that could fit into a metre-square box. This contained the hardware required to support the laser in the railway environment such as a chiller and electrical power-control devices. The new laser system had a higher pulse rate than the original and used fibre optics to deliver the laser beam onto the track. This system was used in further trials. LaserThor also carried out experiments to see whether exposing a railway track to intense laser light damaged it in any way. Sections of rail were exposed to a number of laser shots over a range of intensities and then sent for metallurgical analysis. The results showed no significant damage to the surface of the track under normal conditions.

Although LaserThor's current system is still too big to be fitted to passenger trains in normal service, Higgins envisages it working on 'rail service vehicles' travelling across the entire rail network. At a meeting in 2005 with the chairman and deputy CEO of Network Rail it was agreed that the system was more effective and cheaper to operate than water jetting, and that it had passed all safety tests. The Network Rail executives confidently expected lasers to become the preferred method of cleaning.

The board then changed their minds and said that they would after all stay with water jetting, as the company had invested heavily in it and wanted a return on its investment. As railway companies like to move in the same direction together, no European company has approached LaserThor about using the technology. It appears, then, that rail passengers will have to put up with 'leaves on the line' delays every autumn for the foreseeable future.

Reprieve for the BlackBerry

In the spring of 2006 a deep sigh of relief must have gone round many an executive suite when the BlackBerry, which had been threatened with extinction, survived after a massive payout was made so that it could continue both to be sold and to maintain a service to its existing users.

Research in Motion, or RIM, was founded by two Canadian engineering students, Gary Mousseau and Mihal (Mike) Lazaridis, in 1984. In 1998 they began filing patents for a method by which e-mails are 'pushed' from a host system (such as a personal computer) to a 'mobile communication device' which has a shared electronic address. It was named the BlackBerry for the many small buttons on it which were thought to look like fruit pips. Strawberry was the first suggestion for a trade mark, but 'straw' just sounded too slow.

The main drawing, reproduced here, shows how the system is intended to work. It quickly became addictive for those to whom immediacy of communication was paramount – after all, mobile phones had become so passé, being owned by just about everyone – and soon embarrassment resulted to any supposedly thrusting executive who attended a meeting without one.

Then in 2000 NTP Limited, a Virginia company, sent RIM a letter demanding licensing fees. In November 2001 NTP sued RIM in the Richmond District Court. NTP claimed that five of its patents, beginning with US 5436960, and all with Thomas Campana as one of the inventors, had been infringed. The company had been founded in 1992 by Campana, a University of Illinois-trained electrical engineer, and Donald Stout, his patent attorney, to protect Campana's intellectual property. All the patents published in the company's name give Campana as an inventor. These patents had been filed from 1991 onwards.

The complicated action went on for years, with partial victories for both sides as the struggle produced appeal after appeal. NTP paid its mounting expenses by bringing in over 20 minority investors. In August 2003 an injunction was issued against RIM which would have stopped business for the BlackBerry, but was stayed pending yet another appeal. A 2004 judgment said that NTP's 'particular innovation was to integrate existing electronic mail systems with RF wireless communications networks' rather than to 'disclose a method for composing and sending messages from the RF receiver', suggesting a weakness in their case. In the Spring of 2005 RIM and NTP reached a $450 million settlement but it quickly fell apart. One source suggested that this was because RIM would have wanted its money back if the NTP patents proved to be invalid.

Meanwhile, the US Patent and Trademark Office was indeed looking at the NTP patents to determine whether they had been valid in the first place, or if they genuinely predated the RIM work. One by one they were found to be invalid, though one rejection was under appeal. Some experts claimed that RIM was being favoured by the swift re-examination of the NTP patents, while others asked why they had been granted in the first place. Clearly the same experts could write an entire book on the dispute.

In March 2006 the two parties finally settled out of court. RIM paid NTP $612.5 million in full and final settlement of all claims against RIM, as well as for a perpetual, fully paid-up licence. It applies, of course, only to the United States, which is where the NTP patents are. No money is to be returned if any or all of the NTP patents are declared invalid. RIM had announced that, as of November 2005, the company's assets

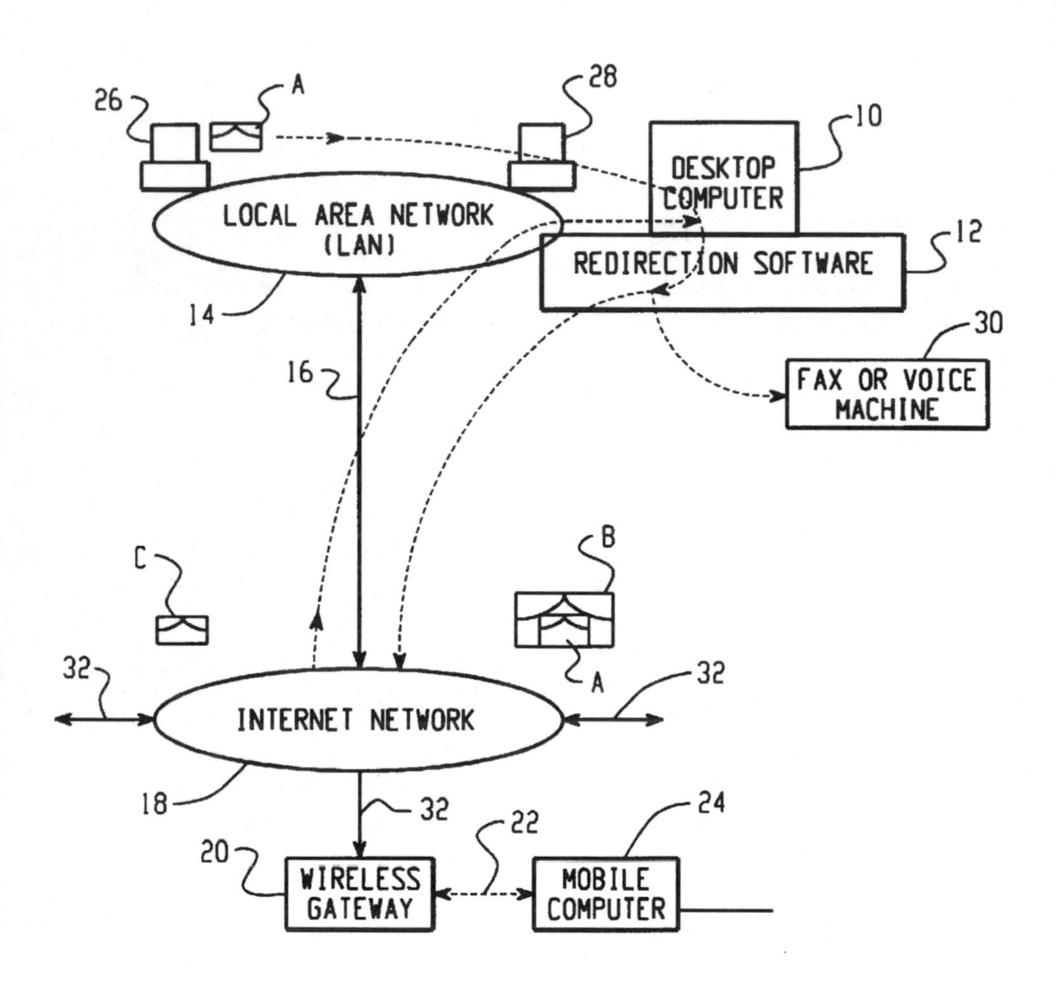
26
A
28
10
DESKTOP COMPUTER
LOCAL AREA NETWORK (LAN)
12
REDIRECTION SOFTWARE
14
30
16
FAX OR VOICE MACHINE
B
C
A
32
32
INTERNET NETWORK
18
32
24
22
20
WIRELESS GATEWAY
MOBILE COMPUTER

comprised $1.8 billion in 'cash, cash equivalents, short-term, long-term investments and escrow funds', so to pay over a third of that amount must have been a major blow. The judge's refusal to delay issuing a BlackBerry shutdown order until the status of the patents was resolved may have encouraged RIM to settle just when things seemed to be going their way.

According to the *Wall Street Journal*, Wiley Rein & Fielding, the law firm acting for NTP, received about a third of the settlement as they were paid on a contingency fee basis. Their turnover doubled as a result, and James Wallace, the lead lawyer, was planning to fight his next case while using a BlackBerry to keep in touch with his colleagues, having acquired it after it was used as a trial exhibit in the RIM case.

Another law firm, Hunton & Williams, who had begun the court action, were (sadly for them) paid on an hourly basis. The rest of the money was divided between a couple of dozen NTP shareholders. Half of them were colleagues of Stout's, the co-founder, at his Virginia patent law firm. Campana himself did not benefit – a heavy smoker, he had died of cancer in June 2004 at the age of 57.

Following the settlement, RIM's share price promptly soared by 17%. The court could have closed down the service altogether, but it hadn't, and the 3.7 million American users of the product could breathe again. Some may wonder how much it would have cost RIM to have carried out a comprehensive prior art search of the patent literature in the first place.

System and method for pushing information from a host system to a mobile data communication device having a shared electronic address, US 6219694, published 17 April 2001

Clean water

There is an increasing shortage of fresh water worldwide. Even where water is available, it is often contaminated or brackish. What is needed is a simple, robust, easy to maintain and preferably portable method of water purification. Besides the developing world, hikers or workers in remote locations could also use the technology. There have been many inventions in this field, several of which are based on similar principles.

One approach to portability was by Dane Mikkel Vestergaard Frandsen and his Vestergaard SA company, with the LifeStraw. It is about 30cm in length and can be carried on a string around the neck.

WO 2008/067817 explains improvements to the concept. There is an anti-microbial surface to the drinking part so that different people can use it. Water is sucked from the source into the unit which contains a 'halogen-based resin' that, it is claimed, kills a wide variety of bacteria. Textile pre-filters remove larger particles and activated carbon removes excessive iodine, bad smell and taste. A new unit removes arsenic, a problem in Bangladesh with subterranean water sources. Chlorine, a waste product in the LifeStraw, is used to oxidise arsenic so that it is dispersed. Water treatment is claimed to be almost instantaneous and the method is cheap, the standard model costing only a few pounds.

In 2007 three Glasgow University students appeared on Britain's *Dragons' Den* TV show requesting £50,000 for their water purification invention, ROSS (Reverse Osmosis Sanitation System). Two were product designers, while the third was a philosophy student. Their invention came out of a final-year project.

A barrel containing dirty water is rolled along the ground from the source to where it is needed and the pressure generated by this movement, they claimed, forces the water through filters, so that the larger contaminant molecules remain in a separate chamber. All five of the Dragons joined to offer the money for a combined 10% stake, much less than the percentage they would usually demand, and the first instance of them all working together.

The students' new company, Red Button Design, claims that ROSS is light enough to be pushed by a malnourished eight-year-old girl in 40°C heat. The filters need changing only once a year. They were quite surprised that no one had thought of the idea before. Sadly, the pressure incurred by just rolling would not, in fact, prove sufficient to force the water through the filter.

Michael Pritchard, a water treatment expert from Ipswich, has combined these two approaches in a portable purification unit which uses reverse osmosis in his Lifesaver bottle, published as WO 2008/037969. It was inspired when he heard about the shortage of drinkable water in areas that had suffered hurricane damage, such as New Orleans in 2005. A carbon filter is combined with reverse osmosis performed by an additional hand pump using air to act on a pressurised water cartridge. Either 4,000 or 6,000 litres of water, depending on the model, can be treated by the cartridge before it has to be replaced. This requirement is simply shown by the water no longer flowing out.

The carbon filter is the dark matter (20) in the drawing on p.60, while the piston for pressurising is in the centre. The liquid reservoir for the pump is at (10). The spout for pouring in dirty water and another for the clean water are at the top and are so placed that opening one seals off the other. A 4,000-litre model costs about £115.

Dean Kamen, well known for inventing the Segway (see p. 124), has also come up with a

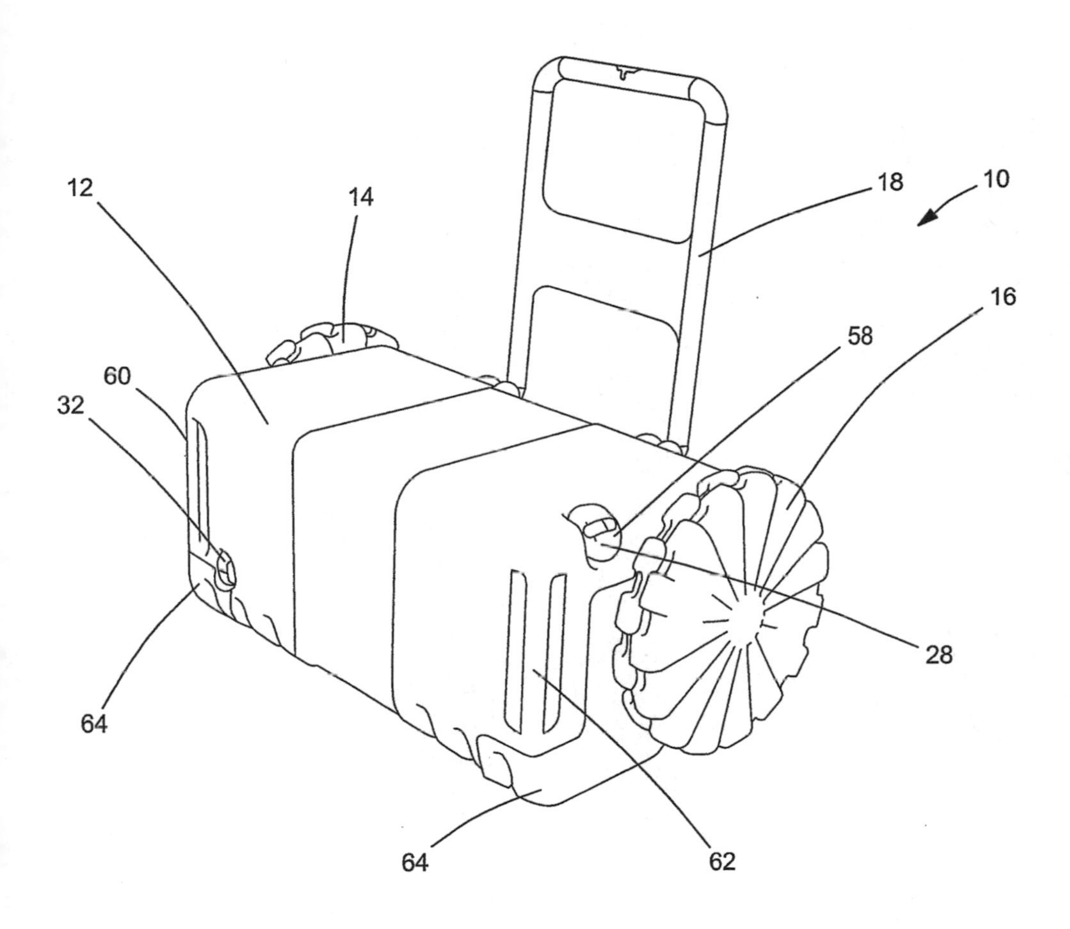

Fluid transport apparatus,
WO 2008/099158, published
21 August 2008

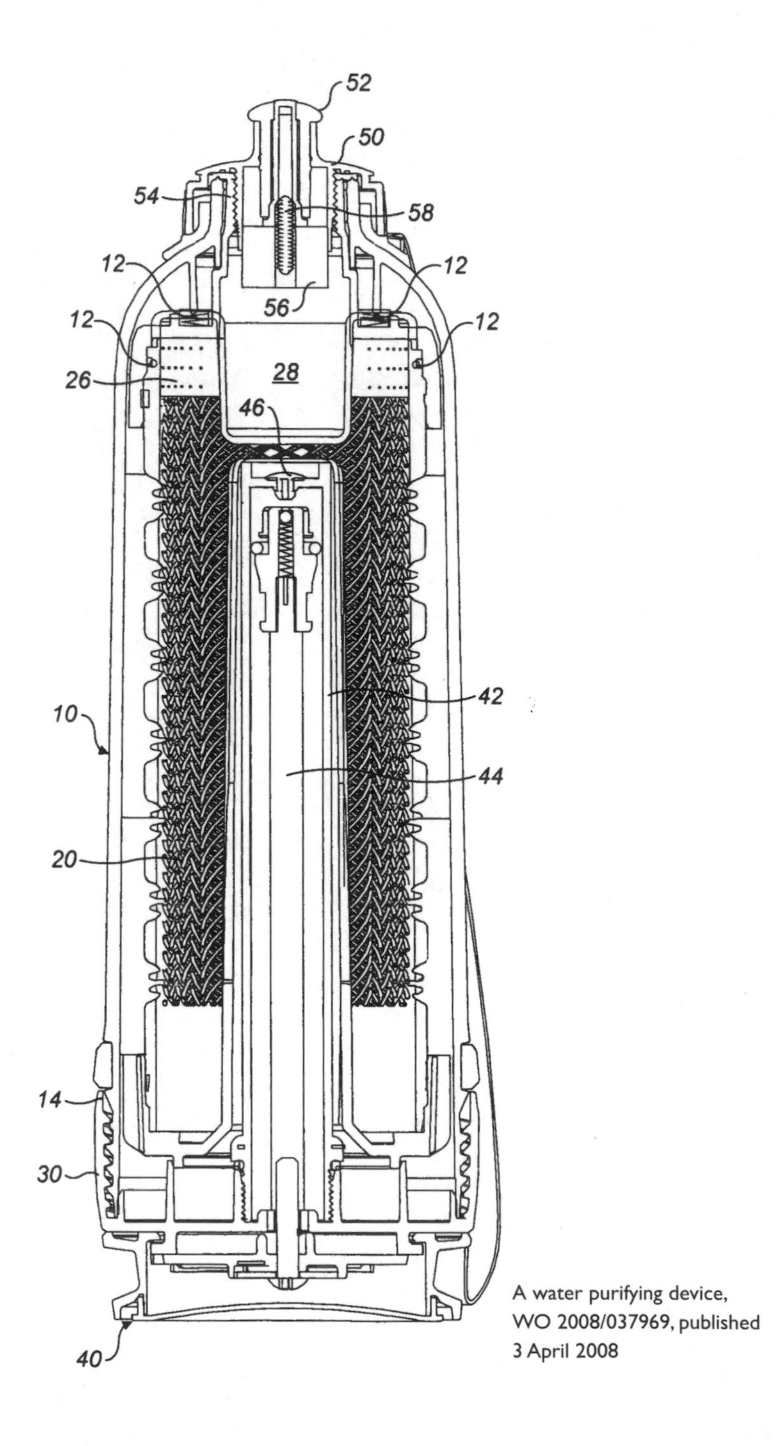

A water purifying device, WO 2008/037969, published 3 April 2008

water purifier aimed at villages in the developing world. Called Slingshot, water is boiled so that it vaporises and then condenses in a second chamber, free from contaminants. This approach is old, but the problem has always been supplying cheap power to provide the heat. Kamen claims that the Slingshot requires only 2% of the power used by conventional distillers. It achieves this efficiency by using waste heat from a Stirling engine that generates electricity. In turn, surplus heat is recycled for the next batch of water using a 'counter-flow heat exchanger'.

The prototype cost $100,000, but the company aims to get the cost of the purifier down to under $2,000. They claim that it would produce 1,000 litres daily while one kilowatt of power, enough to light a small village, would be produced at the same time by the generator (cost, under $4,000). Each machine is shaped like a cube, about a metre in each dimension. The engine itself would run on any source of power, such as cow dung. The invention is explained in US 2004/159536.

Then there is India's Tata Swach, unveiled in December 2009 by Tata Group. It costs 1,000 rupees – about £13. It uses ash from rice milling to filter out bacteria, and also tiny silver particles to kill them. Units are under a metre high, and no electricity is needed. The filter costs about £4 and can treat 3,000 litres, enough for a family for a year, before it closes down automatically.

Future models will be designed to handle arsenic and chlorides as contaminants. Plans are under way to produce three million units annually within five years. The low cost is highly significant for villages in the developing world, but many more are needed for India's population of over one billion.

An earlier version of the product was launched in conjunction with the country's relief efforts following the Boxing Day 2004 Asian tsunami.

Carbon capture in the ocean

There has been increasing talk of reducing global carbon emissions, and about storing, capturing or 'sequestrating' carbon dioxide (hereafter just carbon) already being given off by, for example, fossil-fuelled power stations.

The patent literature reflects this growth of interest. A simple keyword search on the priced Derwent World Patents Index database suggests that during the five-year period 1994–8 there were only 14 patent specifications published on the subject. This grew to 33 in 1999–2003, and then leapt to 144 in 2004–8.

The problem is that there is no established way of carrying out carbon capture. One suggestion is to use carbon to squeeze crude oil out of subsurface reservoirs, so that little is wasted. The carbon would, of course, remain there.

An alternative approach is geo-engineering, whereby large-scale methods are used to alter the climate. Computer models might prove the theory, but many doubt that the methods are truly practical. In addition to feasibility concerns, there are cost considerations and the danger of unforeseen and possibly undesirable side-effects.

Another idea is to store the carbon in the oceans. Over 300 patent documents have been published on the subject. Some are by the Ocean Nourishment Corporation, an 'ethical' Australian company which states that the oceans hold 40,000 gigatonnes of carbon while the atmosphere holds less than 1,000, and that mankind is adding a mere 10 gigatonnes to the atmosphere annually. The chief executive of the company is Professor Ian Jones, who is also director of the Ocean Technology Group at the University of Sydney.

Ocean Nourishment Corporation claim on their website to have patented their technology, although at the time of writing the applications published in their name await grant. The basic idea is to deal both with excess carbon and to provide extra food from the oceans by stimulating the growth of phytoplankton, a much preyed-on food for fish.

About 70% of the oceans lack nitrogen, an essential nutrient for phytoplankton, while nitrogen makes up over 75% of the earth's atmosphere. The company's ideas involve extracting nitrogen from the air and combining it with hydrogen to form liquid ammonia. This is then combined with CO_2 at high pressure to form ammonium carbamate, which is then dehydrated to form urea and water. The liquid urea is mixed with other nutrients. This mixture is diluted in seawater and is delivered, via marine pipelines or ships, to the continental shelf where it passes into the top, sunny zone of the ocean.

This nutrient stimulates the growth of phytoplankton, which also use photosynthesis (which involves taking in carbon). Phytoplankton form the base of the marine food chain and their increased growth both absorbs CO_2 from the atmosphere and stimulates the marine food web. The result mimics the natural upwelling of nutrients from the depths of the ocean.

Phytoplankton live for only about five days. They decompose and either sink to the ocean floor to store the carbon or release further nutrients that support further phytoplankton growth. One of the corporation's patent applications suggests that the best place to feed-in the nutrient would be in a deep ocean such as the boundaries of the Antarctic region.

Referring to the dual benefit of stimulating ocean fish stocks that this method of carbon capture would bring, Professor Jones has said, 'We transform the land to provide food for people. This is like practising agriculture at sea.' At present fishing in the ocean is more like hunting.

In 2007 the company carried out a small experiment in the Philippines using one tonne of nitrogen. There were complaints from Greenpeace saying it would lead to harmful algal blooms, though the company say that the amounts used would be less than 20% of the quantity needed to do that. They estimate that if their approach were adopted then a quarter of current carbon emissions could be absorbed, and that their technology could be used to claim carbon credits under the Kyoto Treaty.

Professor Jones unsuccessfully called for ocean carbon sinks to be included in the Copenhagen Climate Deal in December 2009.

A different approach with a similar effect is by the Plymouth Marine Laboratory in Devon. They propose a floating light collector where power is passed down a fibre optic cable deep into the ocean. An 'artificial light zone' would then diffuse the light, encouraging a phytoplankton bloom, as illustrated below.

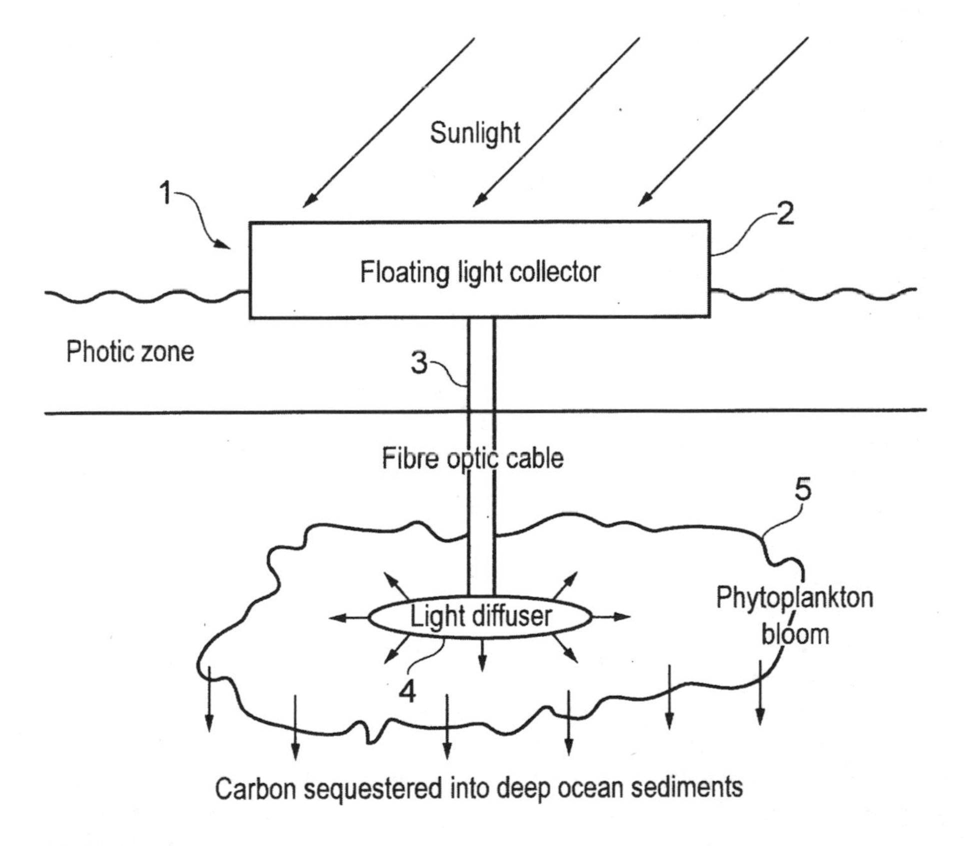

Improvements in or relating to carbon sequestration, WO 2010/029285, published 18 March 2010

Stamping out smoking

Many people cheered 1 July 2007 as a day of freedom from tobacco smoke as England joined the rest of the UK in imposing a smoking ban in enclosed public areas. Others gnashed their teeth at the idea of the nanny state, at losing rights, of interference in others' lives – and, perhaps, at the thought of withdrawal pains.

I am firmly in the first camp, but would like to make a modest contribution to the second by mentioning inventions designed to aid those wishing to give up the weed.

Many inventions consisted of devices that in some way restricted the supply of cigarettes to smokers. The European Classification for patents (ECLA) offers A24F15/00B for cigarette boxes with means for limiting the frequency of smoking, and A24F15/14, cigarette boxes that release a single cigarette at a time. The former class on the esp@cenet database at the time of writing contained 51 American patents, dating back to 1947; the latter, 78, dating back to 1930. The big rush began in the 1960s. This early start may surprise many.

A recent invention is WO 2004/079686 by two private inventors, both called Schaffner, from Florida. It is called 'Moment of impulse anti-smoking message system'.

It is, perhaps, based on the same concept as the cigarette packs that from time to time vary their compulsory warnings of dire consequences. The patent specification begins by discussing US 5228848, which is for a cigarette lighter 'with an anti-smoking message stored on an analog storage chip capable of emitting a short verbal message'. Only one message was provided, though it could be overwritten.

The Schaffners suggested that there were two drawbacks to this approach. First, it was a relatively expensive item. Second, and more importantly, the anti-smoking message was not delivered at the 'critical moment'. The message was uttered only as the smoker was already lighting up, by which time few would have the will power to stub the cigarette out.

'What is needed, then,' the inventors went on, 'is a system for sequentially delivering multiple anti-smoking messages, one at a time, so that the user does not become jaded by hearing the same message and thus does not stop listening to it. A need also exists for a system that delivers an anti-smoking message at the moment of greatest psychological impact – the moment before a fully committed decision to smoke has been made, i.e., before a cigarette has been retrieved from a pack and placed between the lips.'

The act of opening the container in an illuminated environment sets off a circuit and hence triggers the message. In a second version, a normally open switch in the electrical circuit is closed when the lid is removed from the container. This activates the circuit independently of lighting conditions, presumably in case the harassed smoker tries to sneak a cigarette in a darkened room.

A different approach to messaging is taken by another invention, US 2002/0114223, the QT-Watch from Electro-Med Technologies. One of its Illinois inventors is a doctor, the other a former NASA engineer.

A standard wristwatch is adapted so that pushing a special button displays messages, which are 'encouraging, statistical or informational' to help the owner quit the habit. The button is pushed whenever the owner lights up, and the message could, for example, display the number of cigarettes smoked that day, the time the last cigarette was smoked, and an encouraging message.

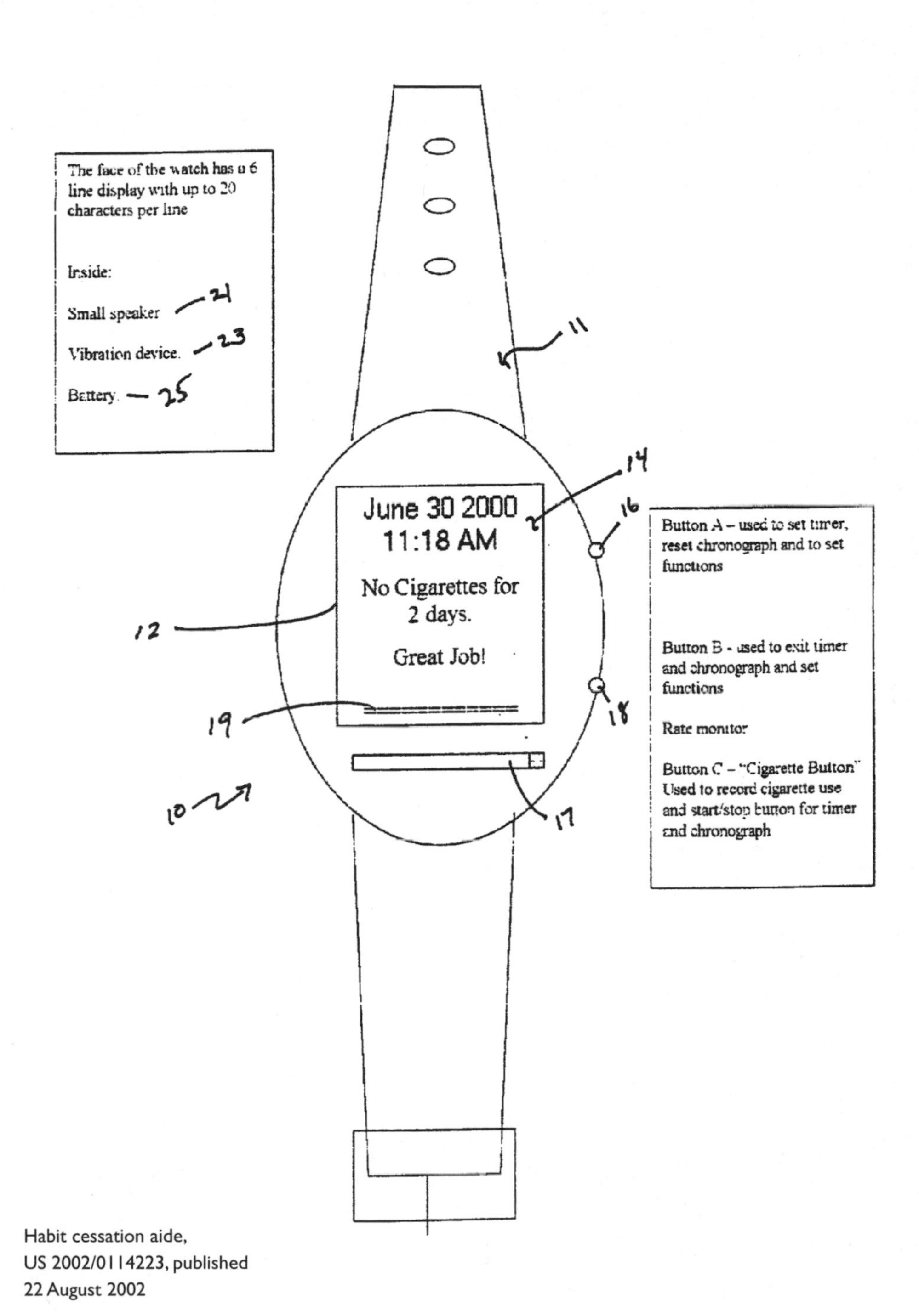

Habit cessation aide,
US 2002/0114223, published
22 August 2002

It sounds quite complicated: the smoker is required to push the button at the beginning of the day and also to set it when lighting up so that a random alarm can sound shortly after the cigarette has been lit to suggest stubbing it out. Messages appear at intervals to keep the abstaining owner's spirits up. Bar graphs can be displayed.

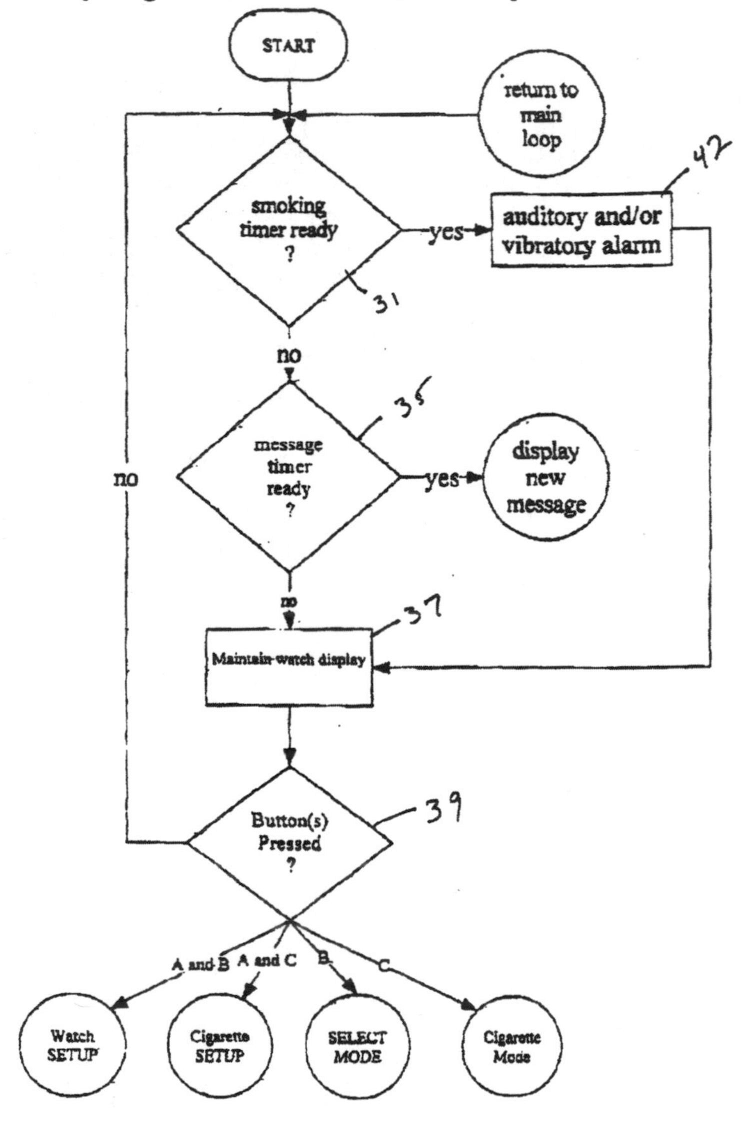

Heelys shoes

Most parents of teenage children will know about the shoes with the wheel in the heel. Their inventor, Roger Adams, grew up in Tacoma, Washington state. His parents owned the biggest skating rink in the Pacific Northwest. Adams, a clinical psychologist, loved tinkering with things like automatic door openers and light sensors. At the time of his invention, however, he was a stressed-out mental-health supervisor who hated his job (he wanted to work directly with people), was going through a painful divorce, and was in the midst of what he acknowledges was a mid-life crisis. One day he was on holiday, watching kids going back and forth along the waterfront on their inline skates. The sight made him nostalgic for his childhood. Then it came to him. 'I had an idea of a shoe that could roll on command by just shifting your body weight,' said Adams. 'It was like a flash; the hair stood up on the back of my neck.'

He cut out the heels of a pair of running shoes and ran a rod through the heel using a wheeled bearing, similar to those used in skateboards. After many modifications, and quite a few falls, he had a practical prototype. He applied for a patent for his 'Heeling apparatus and method'.

Adams's design looks like a normal thick-soled sports shoe, but a detachable wheel can covertly fit in the heel. As the patent's own front-page abstract says, using a single wheel in the heel 'requires a newly learned skillset of balance, positioning, and coordination'. By shifting body weight, wearers can change from strolling to skating, but with no need to carry a skateboard around with them.

Adams knew that a good idea on its own was no use: he needed finance, and then he needed publicity to make the public aware of and excited about his product.

The financial angle was solved, as so often, by contacts. Patrick Hamner first came across the product one evening in March 2000, when he returned to his Dallas home. His sons ran to the door squealing about a video dropped off by a neighbour that showed young people skating on their heels through a parking garage. 'Three boys, yelling, "Daddy, Daddy, this is the coolest thing we've ever seen",' he recalled. Fortunately for Adams, Hamner happened to be a vice-president of a venture-capital company while the neighbour was a colleague of Adams's patent attorney. Two months later, the venture capitalists invested $2.4 million in what became Heeling Sports.

A marketing strategy was devised jointly by a shoe designer, a public-relations firm that specialised in youth markets, and a newly appointed chief executive who had been a manager in a company making roller blades. The product would be aimed at 8- to 14-year-old boys. The plan was to start small and 'stay hip'. Heelys would first appear with little fanfare on the feet of cool kids, go into wide distribution in the spring of 2001 and grow in a swift but 'noncorporate' way. The shoes made their debut at the Action Sports Retailer Trade Expo in San Diego, which features thousands of skate, surf and urban streetwear products. The trade-show director said that the shoes were 'cool and kind of irreverent' because of their ability to switch between ordinary sports shoes and a new, fun sport.

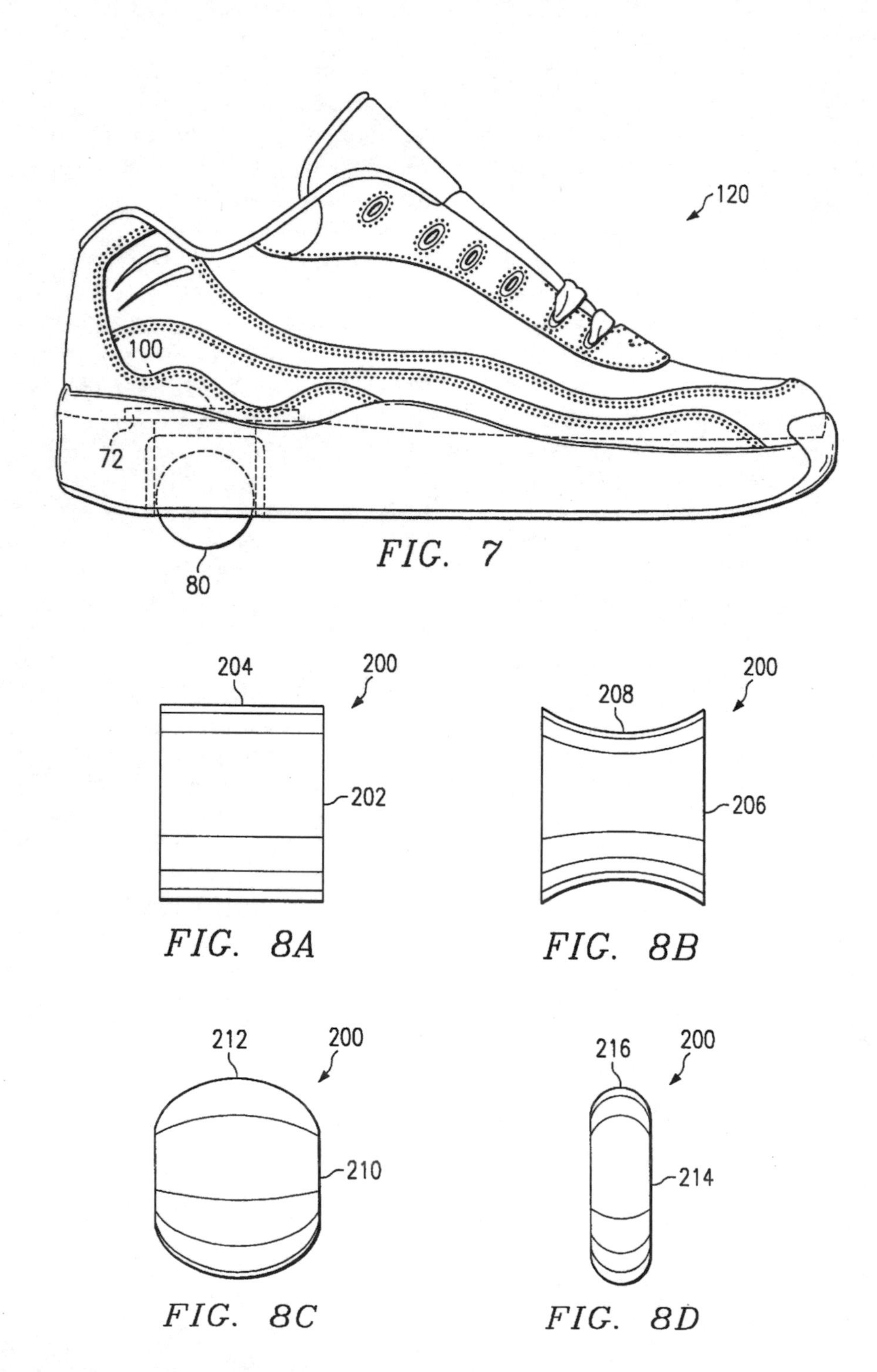

FIG. 7

FIG. 8A

FIG. 8B

FIG. 8C

FIG. 8D

Most major retailers were already selling shoes with wheels in the front and back, so instead the shoe was offered to skate and surf shops plus selected smaller outlets with what they thought was the right image. A website was set up that recruited 40 young people to try out the shoe. They were told to go to shopping malls, skate parks, college campuses and amusement parks at weekends, performing Heelys exhibitions.

Within a few months word-of-mouth publicity had built up. Adams didn't feel this was enough: he wanted Christmas sales. He paid $5 a pair to fly in 7,500 pairs from a manufacturer in South Korea to be sold in ten markets during the run-up to Christmas. Free samples were sent to talk-show hosts and leading television stations and newspapers. Sales began to take off, and by the end of 2001 a million pairs had been shipped. In September 2004 FlyWalk were appointed as the sole British distributor. Their website claims that people wearing Heelys will be noticed, which is probably true so long as the product has rarity value. In the UK, television advertisements have appeared on the Nickelodeon children's TV channel and Heelys have featured on both the BBC's *Dick'n'Dom* show and the national news.

The shoes themselves are made in China and South Korea. Despite the original marketing target, buyers include those in their twenties and thirties (or older), though older wearers do take longer to get used to the new sensation.

Heelys are now sold in more than 30 countries. Sales peaked at over $60 million before falling back sharply in 2009 to about half that figure, and an operating loss, due perhaps as much to market saturation as to the recession.

For Adams, his success validates the way an idea can gain irresistible momentum. 'There's a magic to believing in something,' he says. 'Follow your dreams and roll the dice. There are plenty of doors that are shut; find a way to open one. If I'd listened to the many people saying, "No, you can't do this," I'd have never gotten off square one.'

Heeling apparatus and method, US 2001/0019195, published 6 September 2001

The Silent Knight anti-snoring ring

While visiting my local chemist I came across the Silent Knight finger ring, which, its manufacturers claimed, could prevent snoring. This find was the beginning of the most popular posting ever on my blog. I always like to check a product's packaging for any information on innovations, and there was quite a lot of intellectual property claimed:

Worldwide patents pending
UK Patent Numbers 03285392 and 03201597
US Application Number 10/730.250
Registered Design Numbers: 3015508 and 3015509

Surprisingly, what is almost certainly the most valuable asset to the owners was not claimed. Silent Knight was not at the time a registered trade mark, nor were unregistered trade mark rights claimed by using a ™. Not until later was Silent Knight registered as a European trade mark.

The two inventors, Michael Carter-Smith and Roger Whitaker, are from Surrey. The British patent numbers did not correspond with published documents and I discovered that they were, in fact, the numbers used during the application process. The initial application for the 'Acupressure finger ring' had been filed for and then refiled (hence the second number) before being granted protection under patent number 2405345.

The American number also proved to be a filing number and the patent application has been published, though at the time of writing it still awaits protection. Patent applications do not appear to have been published in any country other than the UK and the USA.

The final two numbers are indeed British registered designs, one of which is illustrated here.

So, how does it work? The ring, which is designed to be worn on the little finger, has a gap in it (presumably to make it easier to put on: the ring is available in four sizes, each covering a range of five jewellers' ring sizes) and a raised section on its internal edge. When in use, this section exerts pressure on the 'meridian line' on the underside of the little finger. According to the 'age-old principle of Acupressure', this will stop the wearer snoring. The design protects that precise look, while the patent can be broader in scope in explaining the function.

Both packaging and website emphasise that the ring is made of hallmarked silver (hence the price tag of nearly £30), though the British patent application states that 'there is little restriction on the material used', provided it is inert, with a metal preferred, so the use of silver does seem rather pointless (and expensive). On its website, the company says that silver is used to ensure it 'lasts and lasts'.

It is also stated on the website that NHS clinical trials are being carried out on the device, and that 'research found that it cut out snoring completely in over 60% of users, reducing it considerably in the remainder'. To their credit, they also cite the British Acupuncture Council's somewhat sceptical finding that 'there is no specific research or proof that [exerting pressure on] this acupuncture point would stop snoring in everyone'. Patents have to be new but there is no requirement that they need to be able to work. This makes the website claim

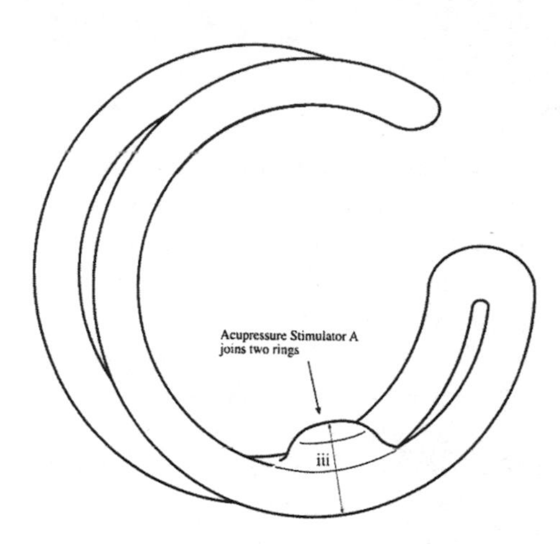

Ring, British design 3015509, registered 24 December 2003

that it is the only stop-snoring ring in Europe with a 'full UK patent' a triumph of style over substance: a patent provides no assurance that the product will be effective. As with many 'alternative' remedies, the ring's effectiveness or otherwise may owe as much to psychosomatic influences as physical ones.

There are a number of testimonials given on the site, including one from a lady who said, 'It is wonderful how it has stopped his snoring but do you have anything for his continuous talking?!'

The hand cream that blocks MRSA transmission

Brian Bennett, a retired tanker driver from Nuneaton, has devised a hand cream which he claims acts as a barrier to MRSA superbugs.

Bennett had relatively little formal education, saying that 'when I left school I was as thick as when I started'. His inspiration came from noticing the pain that his wife Heather suffered every time she washed her hands. She worked in the post office, and handling coins all day gave her dermatitis. The doctors couldn't help, while rubber gloves just made her condition worse.

Wondering whether he could find something to help in the public library, Bennett looked up information about the skin. He says, 'It's amazing what information is available in the library.' From his experience of transporting chemicals he knew that a water-resistant cream was needed which would both moisturise and act as a barrier to bacteria. Silicone turned out to be the right water-resistant agent, the same material used in breast implants. The experts were later surprised that silicone was key.

Bennett asked a manufacturer to make up a water-resistant cream for him. The minimum volume they were prepared to make was 200 litres, which was delivered in a huge drum which had to be moved by three people into his garage. He then mixed 20 litres at a time with other ingredients such as aloe vera, Vitamin E, lanolin, jojoba and evening primrose oil. He kept careful notes of each new solution and how it worked.

Finally, an effective formulation was found: within a minute of rubbing it into her skin, Heather found that it had been completely absorbed. After four days the cuts on her fingers started to close, and within two weeks her hands had completely healed. Not surprisingly, Heather encouraged her husband to get the product into the marketplace.

The idea was partly to form a barrier and partly to offer healing ingredients. It was sold to vets as Vet Shield and to pet shops as Pet Guard. In 2004 a patent application was made by Hygieia Pharmaceuticals. This company was formed specially to exploit the invention as the result of an investment by a recruitment agency. The British patent has been granted, and European and American patents are awaiting grant.

The title is 'A cosmetic or pharmaceutical formulation comprising a silicone compound and water emulsion'. There are over 40 pages giving a lot of detail, and including four specific formulations that can be used.

In February 2009 SkinSure International acquired Hygieia and is now actively marketing a range of SkinSure branded products based on Bennett's work. He is their Technical Director.

Self-service checkouts at the supermarket

For a long time I eyed the self-service checkouts at my local Sainsbury's, but was too nervous to try them out. What if something went wrong? I am not alone, it seems. They may shorten queues but are either loved or hated. A report in 2009 said that 48% of the British dislike them, with 46% saying that they don't scan properly, 39% complaining that they can't use their own 'bags for life', and 13% complaining that they have to do all the work.

The impetus behind them seems to date from US 4787467 by private inventor Neldon Johnson, which dates back to 1987. It is key because over 40 later patents from many companies have had it cited against them as relevant prior art.

The patent explains that to save labour costs, a self-service checkout is a good idea. Previous attempts weighed the goods to prevent fraud, but Johnson's read bar codes as well as producing a bill which then had to be taken to a checkout, reducing the labour savings made with the concept (this variant is not used any more). The shopping trolley rests on a weighbridge. The item taken from the trolley must correspond in weight with what is placed in a bag on a separate pair of scales.

Studies show that many shoppers like using the new technology, sometimes to avoid talking to the till assistant (which often occurs in British supermarkets). A friend of mine likes it because he feels embarrassed if he wants to buy just one or two items. I tend to feel that the computers are likely to go wrong, if only when processing discounted goods. The brusque female voice saying 'Unexpected item in the bagging area, please remove' is also unwelcome. And I do like the personal touch of another person scanning my purchases.

The Johnson patent became the property of International Automated Systems, a research and development corporation. By 2001 at least one supermarket, U-Check in Utah, which IAS used for research purposes, had only automatic checkouts. The shop claimed that its labour costs were a third those of comparable competitors. Rather than employing several checkout operators, only one or two roving staff were required as 'fixers' to sort out problems, and a camera was focused on each lane. The set-up cost was $25–35,000 per lane.

Purely in the interests of research, I did eventually try out the system for the first time at my local supermarket in north London. The self-service checkouts there are of the type where a clear pane of glass faces the user (for scanning) and another pane is on a small table in front of it (which is used for weighing). Goods are transferred from one side to the other for bagging. I was confused at first as there was no obvious indication on the screen about where to scan my loyalty card, or where I should place the items to be scanned, or where vegetables were to be weighed. I could have paid by cash but, nervous of being given the wrong change, I used my debit card.

The transaction probably took twice as long as it would have at an ordinary checkout, in part because it took some time for me to get the hang of using it, but I didn't have to queue and at least I now know how it works. There are probably some losses to the supermarket by fraud or accident – when weighing goods, a cheaper variety could be selected either

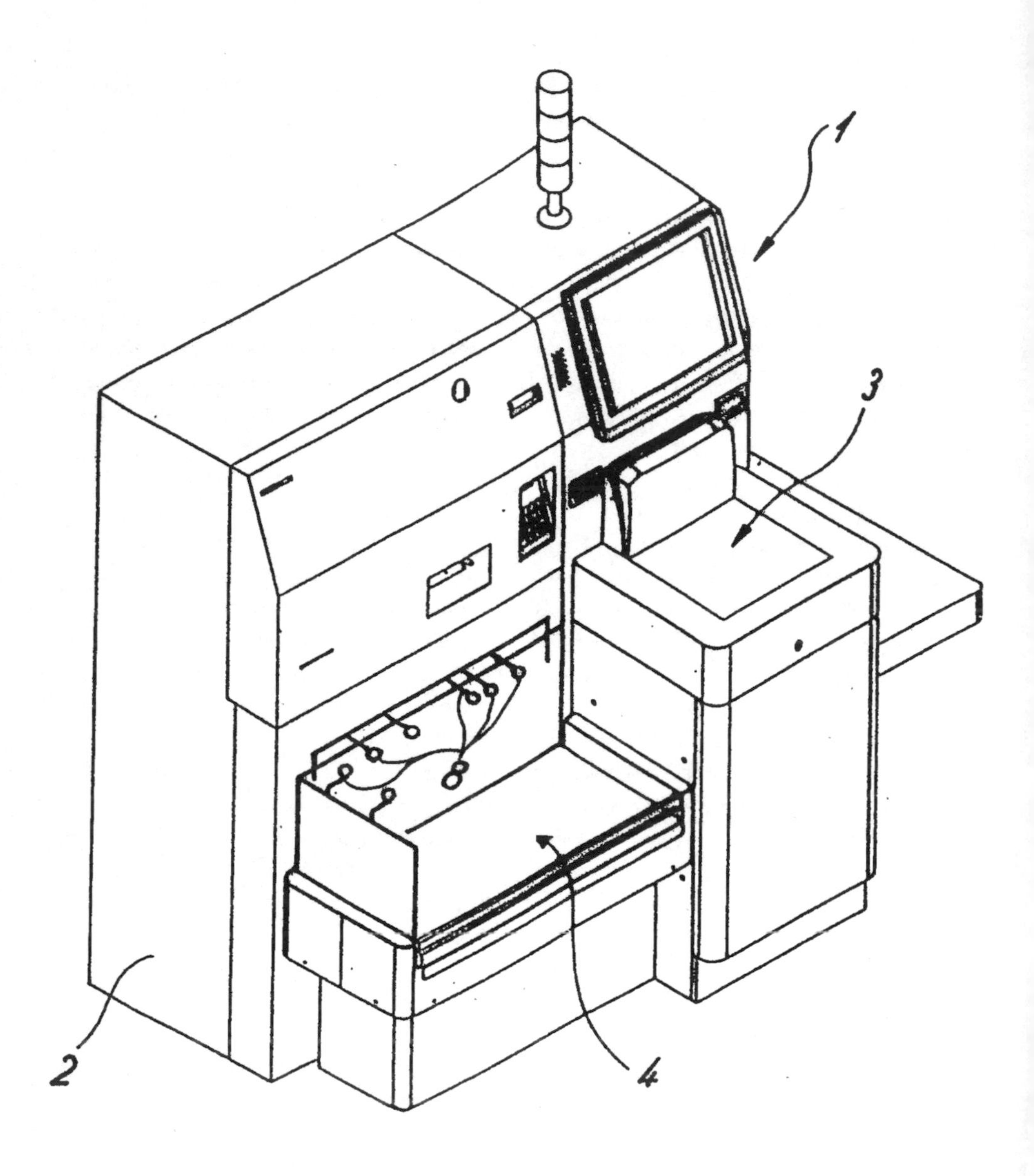

Self-checkout device,
WO 2005/104913,
published 10 November 2005

deliberately or in error from the on-screen options, for example. While looking through patent specifications on the subject, I was amused by the comment in US 2007/241188 that 'While it provides great conveniences to consumers, the online purchase does not provide so much pleasure as doing shopping in real sales points' – not something I was aware of, as I shop out of necessity rather than as therapy.

On the machine I used I could see no information about a patent or model number other than a statement in tiny print that it was made by Epson, but it was fairly similar to the Nixdorf patent drawing illustrated. Patents by prolific companies are often very hard to identify without accurate data.

By 2005 the concept had gained firm acceptance among British supermarkets, and it appears set to expand much more in the next few years. One perceived advantage is saving on floor space, since three or more self-service checkouts can be accommodated in the space taken by a single conventional one, and, of course, fewer staff are needed.

The ironic thing about such technology is that the present generation of consumers is supposed to be better off financially than ever before, yet we increasingly do things that used to be done for us. The days of the petrol-pump attendant are long gone, and we are all now used to refuelling our cars ourselves. Perhaps we will be charged extra in the future for using the services of a human being at a checkout. Or perhaps that option will no longer even be available.

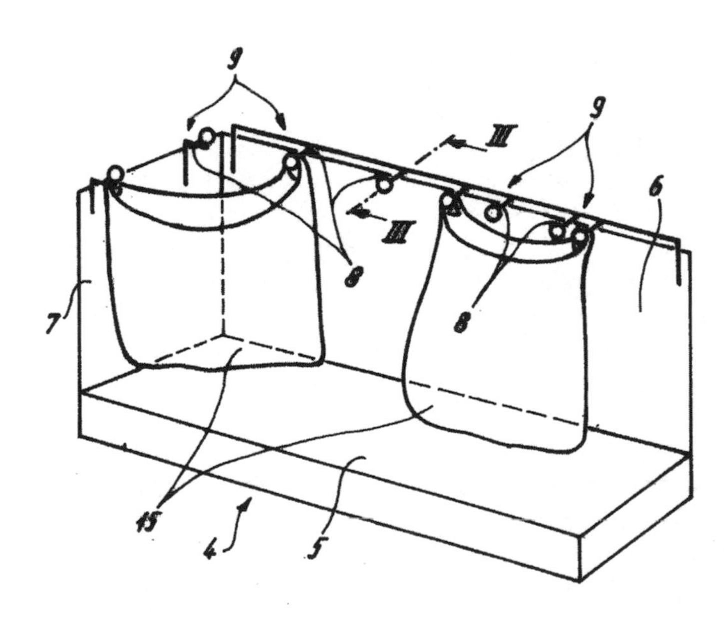

The electronic supermarket trolley

A new kind of electronic trolley may soon be making its way to your supermarket. The idea is to use RFID technology to enable the consumer to get more out of the 'shopping experience'. More realistically, it will result in increased sales for the retailer.

RFID stands for radio-frequency identification, and the technology is increasingly used by retailers to handle goods. At present, bar codes must be scanned by a machine, either at the supermarket checkout or by someone patrolling the aisles. This is labour intensive, so many supermarkets encourage shoppers to do the scanning themselves, either at self-service checkouts or by using a hand-held device that can be taken with them as they shop.

RFID tags on goods offer an alternative. Generally speaking, the technology involves either a passive system, where the user can 'read' data on goods by sending a Wi-Fi message which is replied to, or an active system, where the goods themselves send messages (such as that a food item is past its sell-by date). In theory, then, a supermarket trolley could be fitted with a device that will identify RFID-tagged items as they are placed in it. In practice, the technology will have to become much cheaper in order for usage to be cost effective on any but the most expensive goods.

What the Media-Cart by Media Cart Holdings, a Texan company, proposes is changing the conventional shopping trolley by adding a screen at the front, facing the user. There is also an RFID tag on the trolley itself so that the system knows where the trolley is at any given moment: this is called 'locationing'.

The screen can display advertisements such as special offers for goods that are near the location of the trolley. Offers may be tailored to the known buying patterns of the shopper gathered from loyalty-card data, and onscreen directions provided to special-offer locations. A voice recognition system can respond to consumer questions such as 'Where's the yogurt?' by showing an onscreen plan of the store and the quickest route to the desired item. I suspect that in the UK regional variations in both accent and dialect would make this part of the system work rather less well than the Texan inventors would hope.

Having found the item, the user scans its bar code and its description appears on the screen, perhaps with nutritional information as well. A suggested shopping list could also appear on the screen, again based on previous buying habits.

There is also a 'phone home' facility, enabling abandoned, stolen or otherwise distressed trolleys to 'call in', reporting their location.

All this is culled from the exhaustive detail in the 134 pages of the patent specification.

Adding to the slightly Orwellian feel of the invention, a spokesman for the company says that 'the cart's traffic patterns and shopper interactions are recorded'. The software, says the press release, is by Microsoft, who are collaborating in the innovation.

The summary provided by the company on the front page of the patent specification makes for interesting reading:

> The present disclosure is aimed to address needs of advertisers, retailers, and consumers. Advertisers wish to 1) display ads at the most

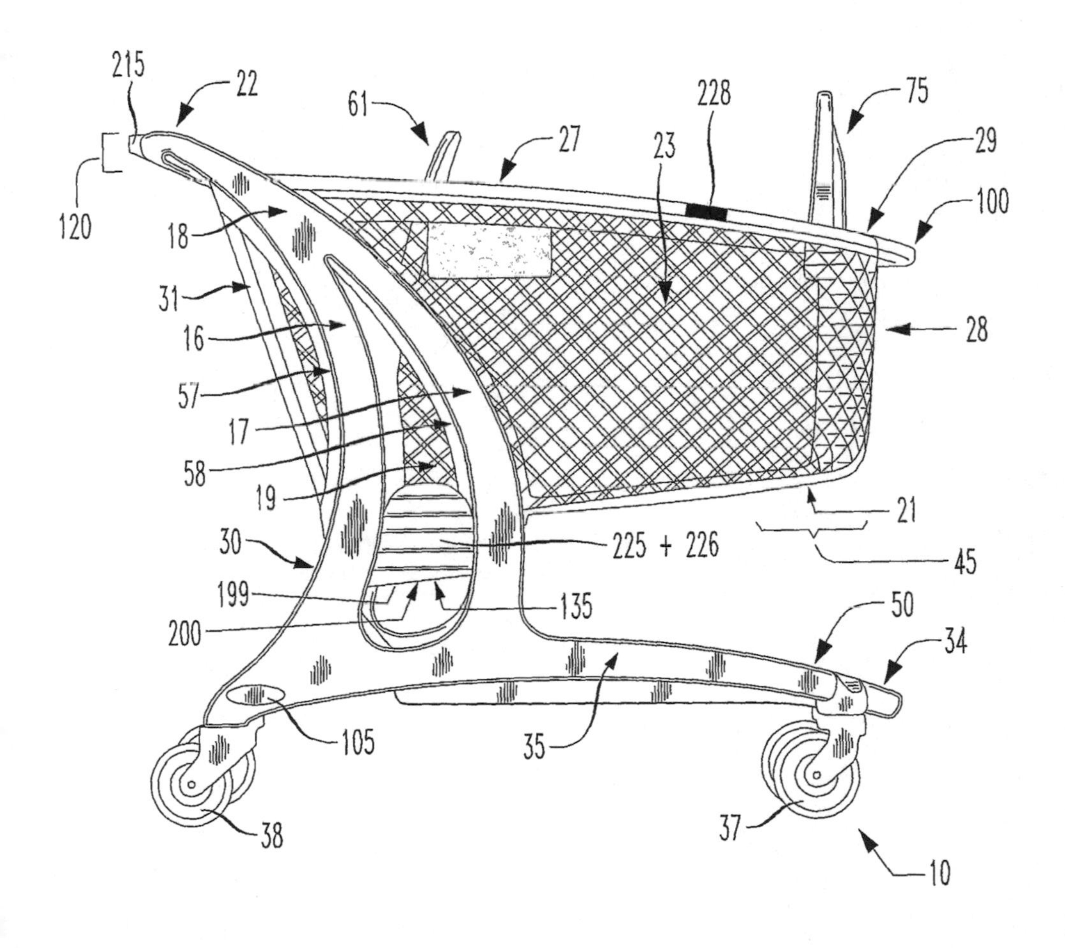

Media-enabled advertising shopping cart system, WO 2007/002941, published 4 January 2007

effective location and time, at the consumer's point of purchase, 2) specifically measure the effectiveness of advertising campaigns, and 3) improve return on advertisement investment dollars. Retailers wish to 1) increase sales, 2) share in advertising revenue, 3) reduce labour costs, 4) create a consumer friendly environment with less advertisement clutter, 5) enhance their store image and 6) make improvements that are compatible with existing solutions. Consumers wish to 1) have a pleasant and efficient shopping experience, and 2) save money on items that they need or want.

A nine-month trial of the system began in early 2007 in the ShopRite stores in New Jersey. One of the appeals to retailers, surely, is the opportunity to recognise people's spending habits (those good old loyalty cards) so that they can be targeted with specific advertising.

Scott Ferris of Microsoft's Advertiser and Publisher Solutions Group is quoted as saying,

> In working with companies like MediaCart, we're continuing to push the envelope in the digital advertising realm to enable new and innovative ways for advertisers and agencies to create brand loyalty and engage with their target audiences in a highly relevant, measurable and targeted way. Digital advertising opportunities are expanding rapidly into new areas, as many of consumers' daily activities, such as shopping, become increasingly 'connected'.

That probably says it all.

Recycling PET plastic

The PET bottle for pressurised drinks was invented by Nathaniel Wyeth of the American chemical giant DuPont. This is the familiar thin-walled, flexible bottle now found in shops throughout the world.

American patent 3733309 was published in 1973 and was basically for the machinery to make the 'biaxially oriented' bottles. The material – polyethylene terephthalate, abbreviated to PET – is a type of polyester, as used in clothing. Wyeth later recalled showing the misshapen results of an early experiment to his laboratory director, who wondered why his company was spending so much money to produce such a 'terrible-looking bottle'. Wyeth was pleased that it was at least hollow. DuPont asked other companies such as packaging giants Owens-Illinois and Continental Can if they wanted to manufacture it under licence, but they and many others initially refused.

Thirty per cent of PET used in manufacturing goes into making plastic bottles, while 60% is used for synthetic fibres. Its important quality is that bottles made from it do not contaminate the contents. Other advantages are that the bottles are light and virtually unbreakable, unlike those made from glass. The material is often also used for sturdy packaging such as oven-ready-meal trays.

Plastic bottles also, however, present a problem with recycling. They take a very long time to degrade and uncrushed empty cylinders take up a lot of landfill space. The first recycling attempts are thought to go back to 1977. Today's bottles commonly feature the triangular-arrow recycling symbol with a 1 in the middle and (in the USA) PETE underneath it. The American Society of Plastics Industry developed the recycling symbols in 1988.

Each number signifies a different type of plastic and enables bottles to be identified by automatic sorting machines. This is vital, as plastics vary in their composition, and the different types need to be sorted correctly. The bottles are broken down into small 'flakes' which are compressed into 'bales' which are then used to make new bottles or other products such as packaging or textile fibres.

According to a recent patent application, WO 2006/020603 by Phoenix Technologies International of Ohio, it has been discovered that there are advantages to making such flakes very small. They 'exhibit unexpectedly superior processing properties for the production of new plastic articles'.

The problem with the tiny flakes, which are easy to decontaminate, is that they are difficult to transport and to handle. The flakes are created in the first place using conventional equipment such as 'grinders, ball mills, impact grinders, cryogenic grinders, pulverisers, attrition mills', suggests the patent application, to produce particles that range in size from 0.012mm to 1.3mm across. These are then heated to the same, or slightly higher, temperature needed to create PET in the first place. This is called the glass transition temperature and is above about 70°C. Rather than forming a hard, solid mass, the heated flakes loosely adhere to each other as pellets. As there are spaces between the flakes, the pellets have similar surface-to-volume ratio as before the heat treatment, and have a superior ratio to traditional methods. This is apparently advantageous in using the material, though the patent does not go into detail. Containers are recommended as a particularly suitable product to be made from PET that has been recycled in this manner, and the material is now available for sale to any companies that want to use it.

Coffee-cup sleeves

Sometimes it is the little item you hardly notice that plays a role in protecting you (or a company's profits).

One summer day in 2008 I was in Massachusetts on holiday and did something that, for me, is very unusual. I walked into a Boston branch of Starbucks and bought a coffee. The cup had a sleeve made of corrugated paper.

I was intrigued to see on the sleeve the mention of three patents: two utility patents and one design patent (prefixed by a D). In American patent law, utility patents cover function and design patents cover the appearance of an item.

Although the sleeve stated an admirable commitment to recycling, I assumed that the sleeve was to protect consumers from scalded hands, and hence the company from being sued. A little research revealed that I was right. In 1995 a Californian woman took Starbucks to court, claiming damages for injuries arising from spilled hot coffee. Such incidents could result from consumers pulling their hands away quickly from a surprisingly hot cup. At the time it was stated that 12 such actions were pending against fast-food chains.

I investigated the patents. The earliest was filed in 1992. It was by the Design By Us Company, of Philadelphia. Its main drawing is illustrated opposite.

The patent is, quite simply, for the corrugated cardboard. It explains that a cheap, environmentally friendly (so not plastic) insulating material is needed 'for thermally spacing the hands of the user from the harsh temperatures of the contents of the container'. I was surprised that the invention was thought to be new – the first patent for corrugated paper goes back to 1872. The 'claims' section, defining the granted monopoly, has wording like 'said tubular member comprises a convex shape along a top edge portion and a concave shape along a bottom edge portion when disposed in said flattened condition'.

More interesting was the other utility patent, which dates back to 2001 and is for a machine to make such insulating sleeves. It has the title 'Beverage container holder', despite being for the machinery to make it.

The applicant for that patent was LBP Manufacturing, Inc., of Cicero, Illinois. In many ways the patents for the machines to make familiar products are more interesting than the patents for the items themselves. It is difficult to invent useful, economically made products, but it is more challenging still to invent machines to manufacture them efficiently.

Sure enough, the abstract provided by the applicant states that the invention is a 'machine and method for producing beverage container holders of consistently high quality at high production rates and at an economical cost'. It is able to produce 50,000 beverage container holders per hour.

The patent refers to the other utility patent listed on the holder and describes its concave and convex edges. It implies that the invention is designed to manufacture this particular holder. The 'blanks' of corrugated cardboard (10 in the diagrams on p.82) are introduced into the machine by an operator in such a way that they are properly aligned, with the correct spacing between them (to prevent jams on the conveyor belts). This is vital as the process is carried out at great speed.

There are many work stations along the length of the machine, at one side or the other, to carry out various processes. These include folding the end flap so that it lies flat over the central section, applying adhesive, the other end

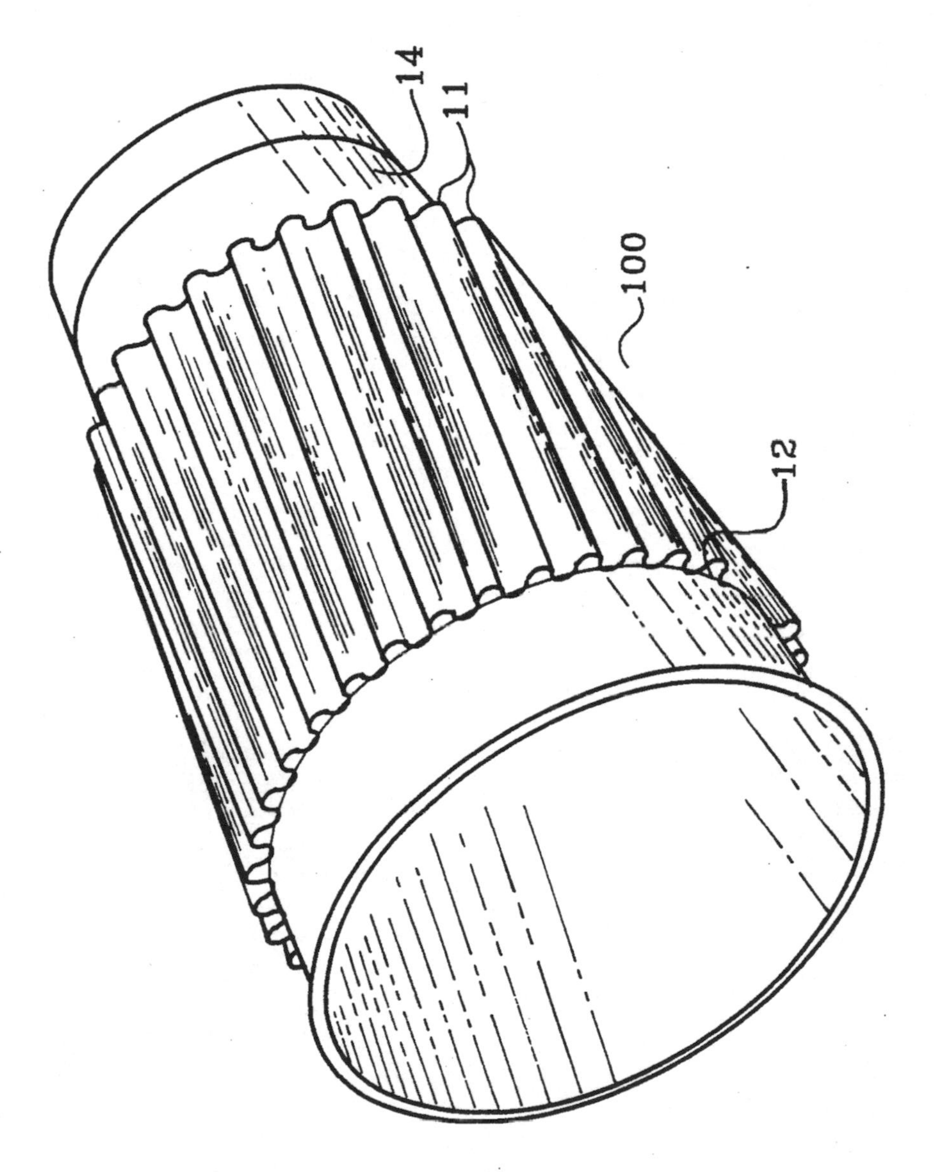

Recyclable corrugated beverage container and holder, US 5205473, published 27 April 1993

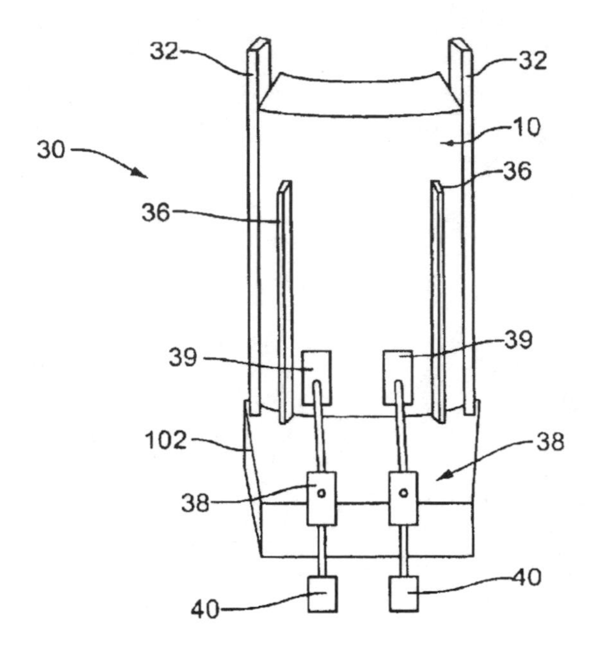
32
32
10
30
36
36
39
39
102
38
38
40
40

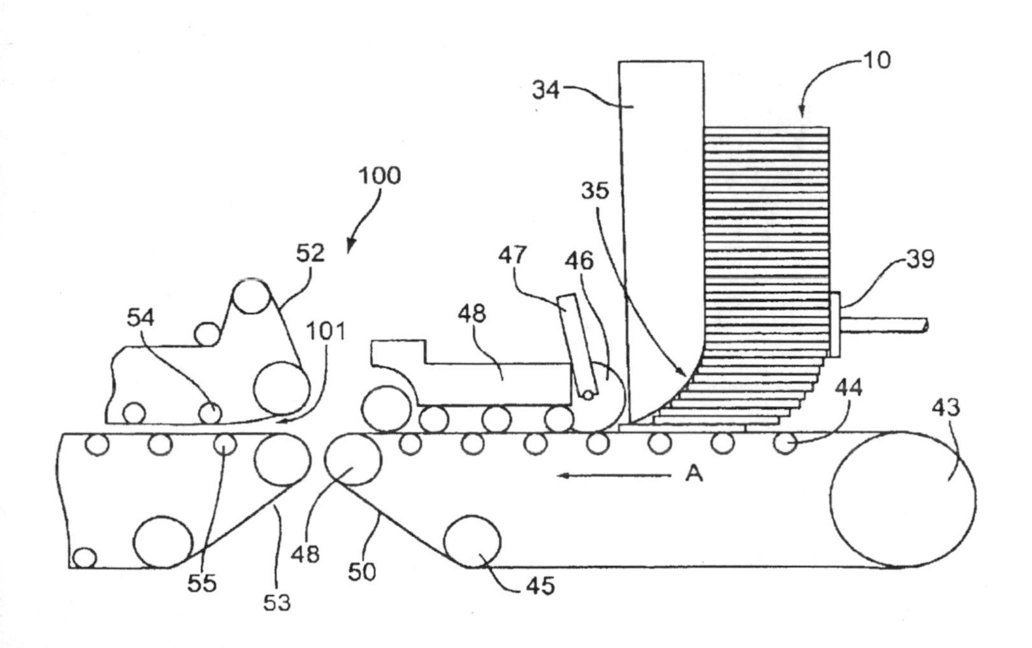
10
34
100
35
52
47
46
39
54
101
48
44
43
A
55
53
48
50
45

flap going over it, and the subsequent application of heat and pressure to activate the adhesive. There is a vast amount of detail in the patent specification about what exactly happens, with an additional 20 drawings. At the end of the assembly line, a second operator loads the holders into shipping cartons.

The design patent is also by LBP and is for the look of the sleeve. Its designer is one of the two inventors of the process patent. All this effort was, and is being, made for an apparently trivial item that is swiftly discarded by customers.

Beverage container holder,
US 6863644, published 8 March 2005

Business method patents

For many years, the US Patent and Trademark Office (USPTO) did not allow the granting of patents for 'methods of doing business'. More and more patent applications for first computer-enabled methods of doing commerce and then business conducted using the Internet emerged from the 1980s. It became increasingly difficult to determine whether a particular computer-implemented invention was a technological or business method. Eventually, the USPTO took the position that examiners would not have to determine whether or not a claimed invention was a method of doing business or technological, as both would be patentable.

This approach was challenged in the 1998 *State Street Bank* v. *Signature Financial Group* court case. This referred to Signature's US 5193056, granted in 1993, 'Data processing system for hub and spoke financial services configuration'. The 'spokes' were mutual funds which pooled their assets in a central 'hub'. The court agreed with the office's practice: this kind of invention is patentable. Ironically, this case is referred to as *State Street*, although that company was opposing the patent.

The impact of one patent is familiar to many. It is 'Consignment nodes', US 5845265, which dates back to 1995. It covers the way the 'Buy it now' function operates on eBay. It was the subject of a court case between its owner, MercExchange LLC, and eBay. In 2008, after six years of litigation, the patent, with some others, was assigned to eBay in return for an undisclosed sum.

The patent talks of 'creating a computerised market for used and collectible goods by use of a plurality of low cost posting terminals and a market maker computer in a legal framework that establishes a bailee relationship and consignment contract with a purchaser of a good at the market maker computer that allows the purchaser to change the price of the good'.

Machine code is not required (let alone given) in patent applications, only the details behind the flow chart outlining what happens and where decisions are made. Such a patent would mean an entire procedure is blocked, rather than particular methods of carrying out the procedure, which fall under copyright law. For example, in a dispute that lasted from 1999 to 2002 Amazon's patenting of a 'one-click ordering' system enabled them to force Barnes & Noble to modify their own online procedure for existing customers to a method requiring two mouse clicks rather than one. The position in Europe is complex, but basically software patent applications are much more rarely granted than in the US because in Europe administrative (rather than technical) solutions to problems are not regarded as inventions.

One consequence of the rise of business method patents was the need to create a new class in the American patent classification. Class 705 was created in 1997 by combining two older classes. Its title is 'Data processing: financial business practice, management, or cost/price 'determination'.

An official study found that in that class the number of postal meter inventions was surpassed in 1995 by financial transaction systems, and in 2000 by electronic shopping systems. To date, over 20,000 Class 705 patents have been granted.

However, over 55,000 patent applications are in that class, and while some have no doubt been rejected many must still be pending. The delays are partly due to the difficulty in finding examiners with financial services backgrounds who are willing to accept the USPTO's relatively low pay scale. Patent examiners in computer engineering start at $52,000 per annum, while most examiners start at $40,000. The 6,000 patent examiners have only 20 hours of work time allotted per application. This is used to search for prior art and to decide whether a specific application should be rejected, accepted or modified. There is a steady drain as many examiners are hired by better-paying companies in the private sector, hoping to use their skills as poachers rather than gamekeepers. A million applications in all fields of utility patents are awaiting a decision – and the annual number granted has never exceeded 174,000.

Much prior art in this field is not in the form of patent documents. Applicants to the US office are required to list known relevant prior art, but that is often insufficient to enable a decision to be reached. In July 2008 the USPTO launched a two-year pilot 'Peer to Patent' programme. Volunteer patent applicants have their applications examined immediately. Meanwhile, interested members of the public review the applications and provide prior art and commentary to assist the patent examiners. Participants are asked to rate the prior art submitted by fellow participants. At the end of the review period the ten highest-rated citations are submitted to the relevant patent examiner.

A year into the trial it was found that, on average, almost five prior art citations were provided by reviewers for each application. Of these, 55% were non-patent, while of those mentioned by the inventors only 14% were. Clearly a lot of relevant prior art, not in the patents, was coming to light for examiners to consider.

Medical advances

It is likely that the coming decade will bring us massive advances in medical diagnoses and surgery, if only we can afford them.

There is, for example, the concept of a smart, or intelligent, pill. Proteus Biomedical is a Californian company which has applied for several patents whose titles include 'In-body device having a multi-directional transmitter' and 'Multi-mode communication ingestible event markers and systems'.

The pill is swallowed and while in the body is capable of sending a signal to a chip which is either embedded below the skin or incorporated into a skin patch. This in turn sends a signal to a smart phone, or perhaps to a doctor via the Internet. The signal could simply confirm that the patient has taken a prescribed medication, or could indicate that adverse reactions were occurring. An 'identifier' would be needed which would initiate the signal when the pill reached the target area of the body.

One example might be the monitoring of a patient's stomach. The presence of stomach acids would act as an electrolyte, which would interact with an anode and cathode in the pill to activate a signal. In January 2010 Swiss pharmaceutical company Novartis signed an agreement to work with Proteus to develop products based on this technology.

Remote sensing, whereby a transmitter implanted in a distant patient can send data, will probably grow as a topic. The first drawing refers to a currently available system, by ineedmd.com, Inc.

The disposable glove (12) contains ten embedded electrodes together with printed circuits. A protective layer is peeled off and the glove is then placed in the correct position on the patient. An adhesive ensures that it does not slip off. Data such as heart rate, blood pressure, and temperature are sent to a laptop computer, where they are analysed by the company's software. The device has been approved by the US Food and Drug Administration (FDA), and is marketed as The Physician's Hand.

Robotic methods are increasingly being used for surgery and invasive examinations. An example is the 'Modularity system for computer assisted surgery' by Computer Motion, Inc., of California. The company is said to be the leader in medical robots.

Coronary blockages, for example, can be dealt with by robot arms wielding instruments through incisions in the chest. An additional arm uses an endoscope to provide images of what is happening inside the heart chambers. Reassuringly, a flesh-and-blood surgeon controls the robot arms remotely, the surgeon's own hand and arm movements being mirrored by the robot's, and is able to view events onscreen. The robot arms are sold under the name of Aesop and the earliest models date back to 1994.

Tele-diagnostic device, US 2005/0075541, published 7 April 2005

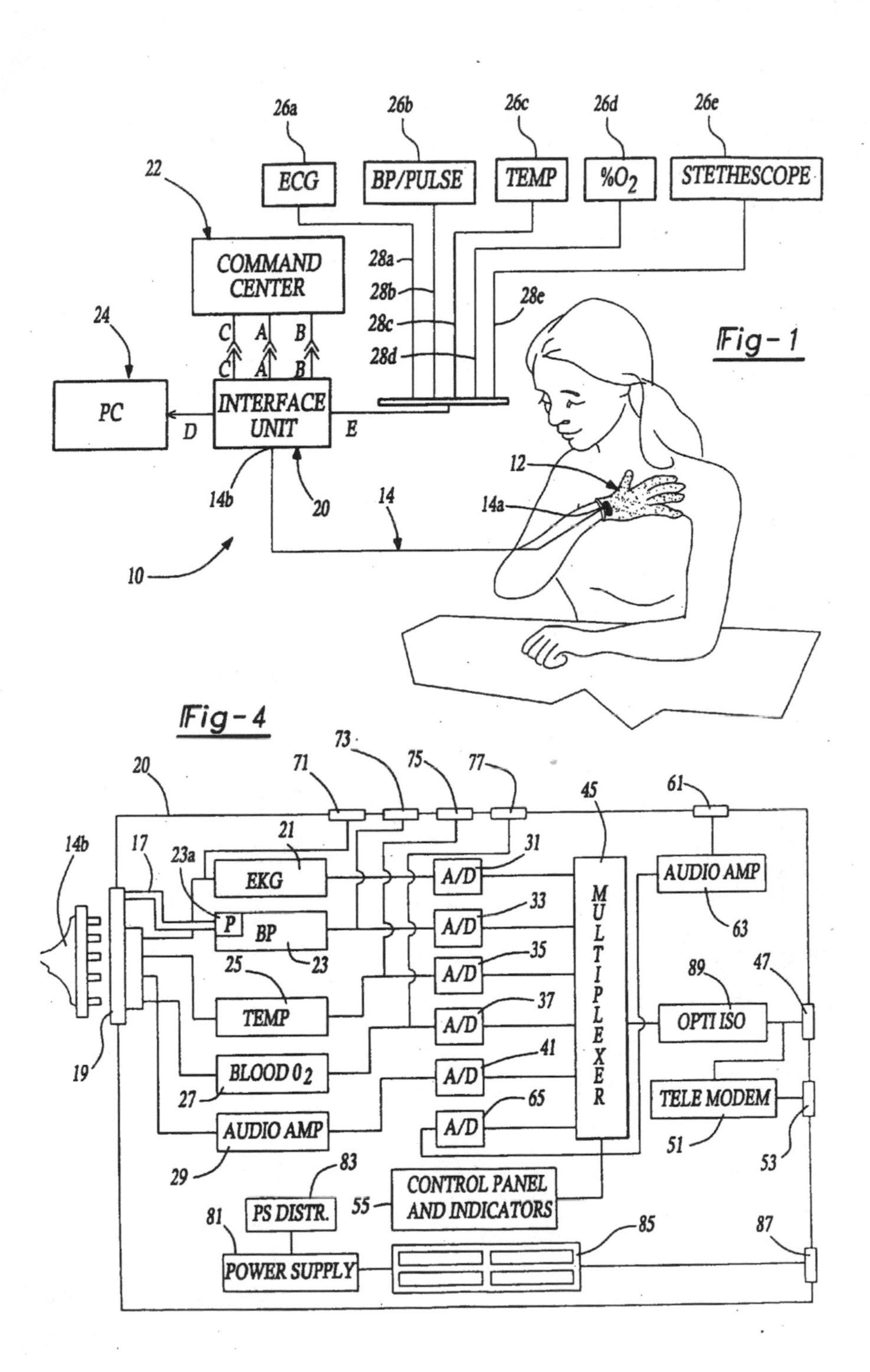
26a
26b
26c
26d
26e
22
ECG
BP/PULSE
TEMP
%O2
STETHESCOPE
COMMAND CENTER
28a
28b
28c
28d
28e
24
C A B
PC
D
INTERFACE UNIT
E
Fig-1
12
14b
20
14
14a
10
Fig-4
20
71
73
75
77
45
61
14b
17
23a
21
EKG
31
A/D
AUDIO AMP
P
BP
33
M U L T I P L E X E R
63
25
23
35
89
47
TEMP
37
OPTI ISO
19
BLOOD 0 2
41
TELE MODEM
27
65
AUDIO AMP
51
29
83
53
PS DISTR.
55
CONTROL PANEL AND INDICATORS
81
85
87
POWER SUPPLY

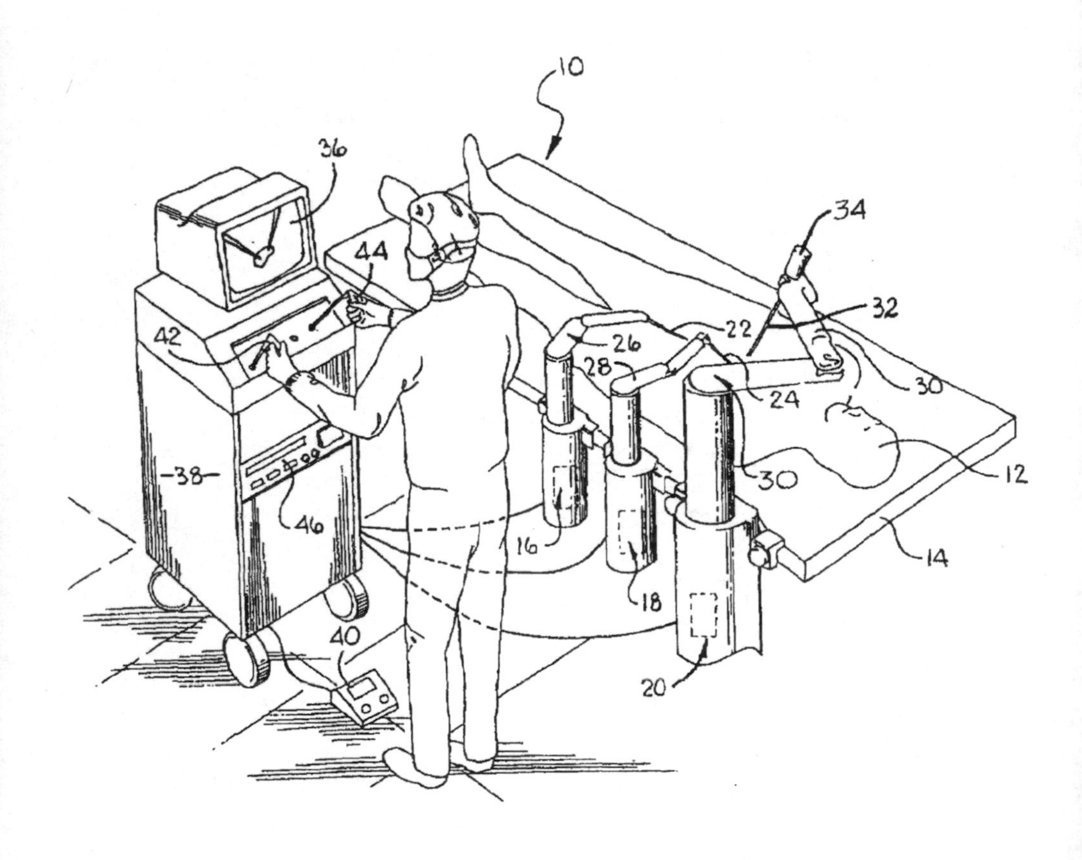

Modularity system for computer assisted surgery, US 2003/050733, published 13 March 2003

Robot helpers

Honda has been working on the Asimo project since 1986, with the slogan 'designed to help people' on its website. The Asimo is meant to be, quite literally, a robot helper.

Asimo has gradually been refined so that it can move in a rather cautious, always upright position. It has been pointed out that the Daleks in *Doctor Who* could easily have been defeated by putting a few steps between them and their quarry, so Asimo is definitely an advance since it is the first robot capable of negotiating a staircase. There have been television advertisements to publicise the company showing Asimo, who looks rather like a short astronaut (it is 130cm high), moving about a stage. The astronaut idea is clever. The facial area is obscured by a visor, so there is no need to provide a visage (or eyes), and the battery pack on the robot's back does not look out of place.

It might be thought that Asimo was named for the science-fiction writer Isaac Asimov and his laws of robotics, which include robots being prohibited from doing harm or giving orders to humans. More prosaically, however, Honda say that Asimo stands for Advanced Step in Innovative Mobility.

A lot of work has gone into evolving this android, and it is the subject of over 100 international patent applications with titles like 'Legged mobile robot and program control'. It is tall enough to reach light switches and so on, yet not so big that materials are wasted, or that it looks intimidating. Powered by a lithium battery, it has a gyroscope and an acceleration sensor. Its fingers can grip, but not so hard as to injure a human. Many of the innovations involve controlling its gait and to deal with floor-landing shock, and there is also software to control how it responds in a given situation, as there is no remote monitoring of its behaviour. Probably the biggest challenge is programming its response to an unusual or dangerous situation – sometimes 'do nothing' is the best answer, though not if a fire menaces humans. The illustrations show how Asimo was portrayed in a recent Japanese-language patent application. The subject of the application is controlling the posture while interacting with a 'target object' (a human being).

Besides its ability to move, Asimo has capabilities which can be listed in five categories.

The first is using a camera in its head so that it can assess the direction and distance of moving objects. Hence it can follow someone, or greet an approaching human.

The second is the ability to interpret the position or movement of a hand so that it can recognise posture and gestures. It will readily shake hands, or will wave back in greeting.

The third is the ability to recognise its environment and take steps to make it safe for itself and for humans. For example, it will avoid hazards such as moving objects.

The fourth is the ability to distinguish sounds and to react accordingly. It will face someone speaking to it, respond accordingly, and will respond to questions, perhaps with a nod or by shaking its head rather than speaking.

The fifth is the ability to recognise faces, even when it or the face is moving. Up to ten different faces can be recognised by a single unit and addressed by name.

Some may wonder at the commercial viability of Asimo. Demographics is key. Japan has the fastest-ageing population in the world. The country's birth rate is very low, and its citizens enjoy a long life span. Over a quarter of Japan's population is already over 65, and by 2050 it will be a third. Honda envisage that Asimo will be not just a companion but also a helper, as there

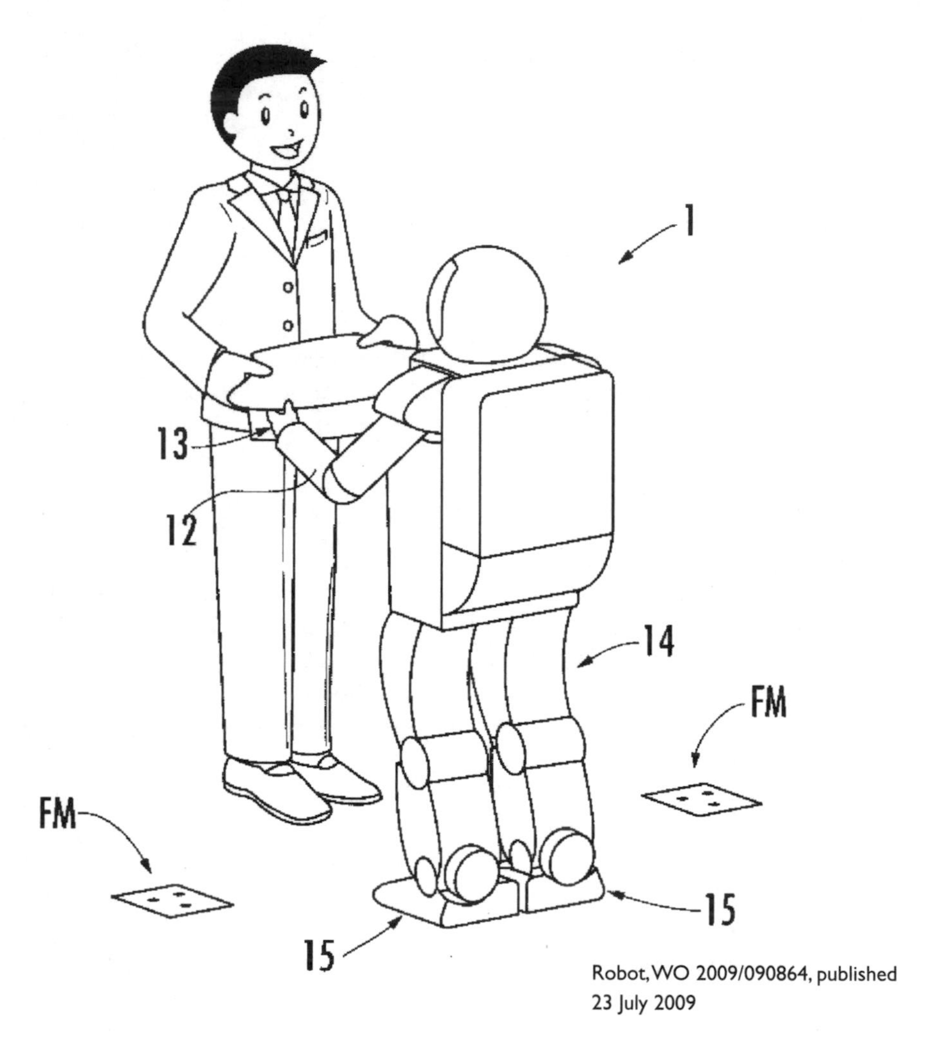

Robot, WO 2009/090864, published 23 July 2009

simply won't be enough younger humans around to assist. The same is true of many Western societies. Bill Gates has predicted that by 2025 every home will have a robot (though as each experimental model currently costs over $1 million to produce, Gates is one of the few people who could presently afford one). Besides helping the elderly with everyday tasks they could be used for nursing, education, housework and security.

The mannered, precise way in which Asimo responds to a given situation or command may seem irritating to Westerners, but would be well received in Japanese society, where a predictable response is always expected in any given situation. Perhaps the biggest challenge for Honda will be persuading people to accept that a highly capable robot can act as a responsible carer for aged relatives.

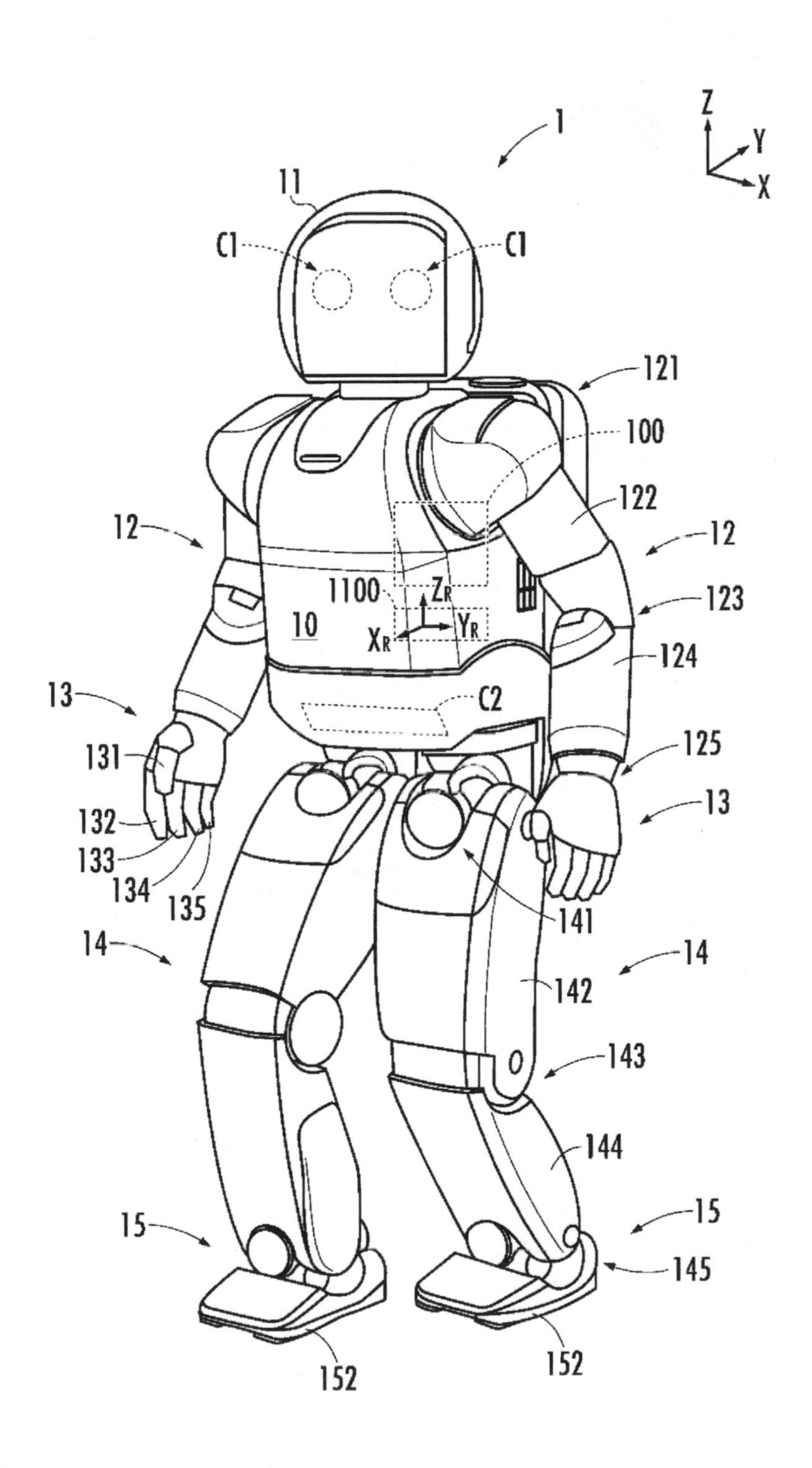
1
Z
Y
X
11
C1
C1
121
100
122
12
12
1100
Z_R
Y_R
X_R
10
123
124
13
C2
125
131
13
132
133
134
135
141
14
14
142
143
144
15
15
145
152
152

Protecting the computer and its users

There are at least 30,000 patent specifications for protecting computers against viruses and other malicious programs. The range of patenting companies is vast, and specifications include dozens by well-known Internet security providers such as McAfee and Symantec (the company behind the Norton brand). Huge potential losses are involved, so computer security is an obvious area of interest to software manufacturers. 'Spam' or junk e-mails waste a great deal of time and may spread many harmful viruses.

Norton and many others protect their own software against piracy using an internally generated alphanumeric code to identify the host computer's configuration, which ties in with the product key.

Most patents in the field employ computer jargon such as 'malware' (malicious software), 'whitelists' (e-mail addresses considered to be trouble-free), and 'blacklists' (the exact opposite).

An example is Microsoft's US 2006/095971, 'Efficient whitelisting of user-modifiable files'. When a file arrives, a whitelist is consulted to obtain a 'trust level', and a decision is made on whether to treat the file as malware. This may sound very simple, but it relies on a sensitive and unique 'signature' for a file so that even a subtle change can be detected. The document explains that to avoid being blacklisted, malware creators sometimes randomly change a portion of a malicious program to avoid detection. Positive verification (appearing on a whitelist) is therefore superior to non-positive one (not appearing on a blacklist).

Blacklisting is a common anti-virus (AV) technique, but most AV programs also use heuristic algorithms to detect suspicious patterns. Heuristics is a problem-solving technique in which the 'most appropriate solution of several found by alternative methods is selected at successive stages of a program for use in the next step of the program'. An AV program might therefore flag a file as suspicious if it exhibits similarities to a program on its blacklist.

A Trojan Horse is a malicious program that does not replicate itself through the computer system (that would be a virus), but rather enables control of the computer by a third party. Trojan Horses often masquerade as useful programs that unwary computer users are fooled into downloading. Once present, they can, for example, send e-mails without the owner's knowledge as part of a 'botnet', steal passwords or log keystrokes. One Internet security provider, BitDefender, estimated in 2009 that Trojans accounted for 83% of malware.

One anti-Trojan initiative has been put forward by Chinese company Inventec, with their Chinese patent application CN 101388059. As far as I can discern from the drawings and a brief English summary, it uses an onscreen keyboard to scramble the keystrokes. Before typing a letter, say a 'c', the user points the cursor at the onscreen 'c' and is given an alternative letter to type, say 'b'. This means the Trojan software logs 'b' when the correct letter is 'c', so cannot correctly identify the words being typed.

Turning to spam filters, there are several thousand patent specifications. Many of these also employ the blacklist/whitelist concepts, where, typically, a query is generated if an e-mail

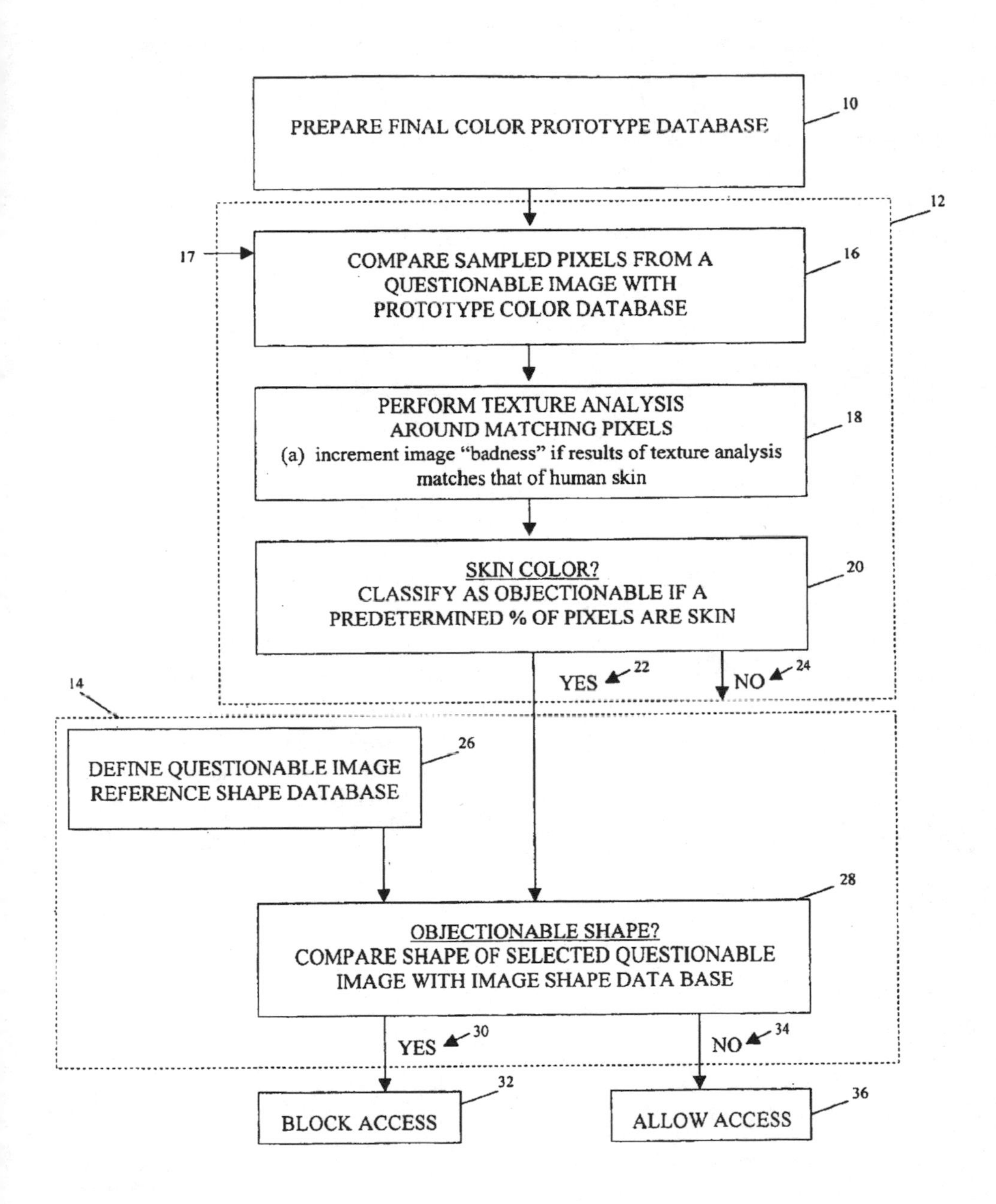

Automated detection of pornographic images, US 6751348, published 15 June 2004

arrives from an address that is not on the user's whitelist. US 6493007, granted in 2002 to a private inventor, has the spammed victims fighting back. A message is returned displaying a 'remove' icon. As many spammers simply want to verify that receiving e-mail addresses are live (so that they can sell them on) this may not be the best approach.

I like the title of the 'Spam avenger', US 2003/009698, by CascadeZone of Washington state. The invention does not allow e-mails from locations that do not respond to confirming e-mails, as the specification claims that most spammers cannot receive messages. Unfortunately, some companies and organisations routinely use 'no reply' e-mail messages when they mail their customers.

An example of building blocking filters using the known attributes of spam can be found in US 2002/199095 by Tumbleweed Communications of California. Negative ratings are assigned to certain words. If say 90% of messages mentioning Viagra are spam then a 90% negative rating is assigned. Users improve the database by nominating specific messages as spam. To avoid losing a genuine message, a mention in 'good' messages is doubled. It is claimed that genuine e-mails are hardly ever lost while nearly all the spam is filtered out.

Finally, a very specific form of protection is an invention by Fotonation Holdings, of New Hampshire. The inventors are ethnically diverse, comprising three Romanians, two Irishmen, one Frenchman and one American.

Pornographic images are detected using a 'color reference database' which contains colours of portions of the human body. Sampled pixels are compared with the database, and those that match are subjected to a texture analysis to find if other comparable pixels surround it. If a large area of possible skin is found, the image is classified as objectionable. These can then be passed to a database of objectionable shapes such as an 'erotic pose'. The result is that possible candidates for screening are manually checked, with a greater frequency if the image is flagged as suspicious, presumably for inclusion in a list of blacklisted sites, or so that the need for 'parental guidance' can be indicated.

Harry Potter and intellectual property

Few Harry Potter fans ever think about the formidable investment in intellectual property involved. It is often said that the real money in the film industry is made from the 'character merchandising' rights rather than the films themselves (or the books that inspire them).

The films are, of course, based on the copyright in the plots and characters which Potter's creator, J.K. Rowling, sold to Warner Bros. In addition, Warner Bros. Worldwide Consumer Products secured a worldwide licensing and merchandising rights agreement as part of the deal. In turn, in February 2000 Warner Bros. licensed Mattel as the Harry Potter 'worldwide master toy licensee'. Other rights such as electronic games were sold to Hasbro.

'All of Mattel is inspired by J.K. Rowling's rich, magical world of Harry Potter and we're thrilled about the relationship,' gushed Adrienne Fontanella, president of Mattel's Girls Products division. Financial figures to support her enthusiasm are thin on the ground, although the sums involved are undoubtedly huge. Databases can be searched to find at least some evidence of this potential goldmine, even if not all the proposed toys make it to the product stage. Searching by company name sounds easy, but the problem is to identify those which are relevant. Mattel has thousands of patents and designs.

A useful but by no means infallible way to unearth relevant patents is to search entire specifications on a database such as Google Patents for a specific term, in this case 'Harry Potter'. Over 200 can be found with such titles as 'Animated photographs', 'Levitating ball toy', 'Interactive dark ride', and 'Game with multiple chambers'. The success of this method is dependent on the applicant mentioning the inspiration in the text, of course, and works better with utility patents (which cover function) than design patents (which cover appearance), since design patents usually comprise only a title and associated illustrations.

All the relevant Mattel patents I was able to locate listed Jeannie Hardie Burns, James Harrison and/or Jonathan Bedford as inventor(s). One idea, shown here, is a box that opens like a book to reveal a gaming surface designed by Bedford.

It is clearly an attempt (rather feeble, in my view) to simulate Quidditch, an aerial ball game that features in many of Harry's adventures. Opening the box causes supports to lock into place so that an upper deck on which the players move is raised above the 'ground'. Pieces representing the Snitch and other characters are shown in the drawing. Actions are determined using dice. This is definitely one case where an electronic version would have been far more interesting.

Then there is US 6932342, dating back to 2001, by Harrison and Burns. It is for collecting items from specific shops in Diagon Alley – Rowling's fictional London shopping street. After setting off from the Leaky Cauldron inn, players have to purchase a number of items such as wands, potion bottles, owls, books, cauldrons and robes. First back to the Leaky Cauldron wins. Along the way Spell and Havoc cards can be used and 'Wizard money' is collected each time Gringotts is passed. This game, at least, can be purchased as the Harry Potter Diagon Alley

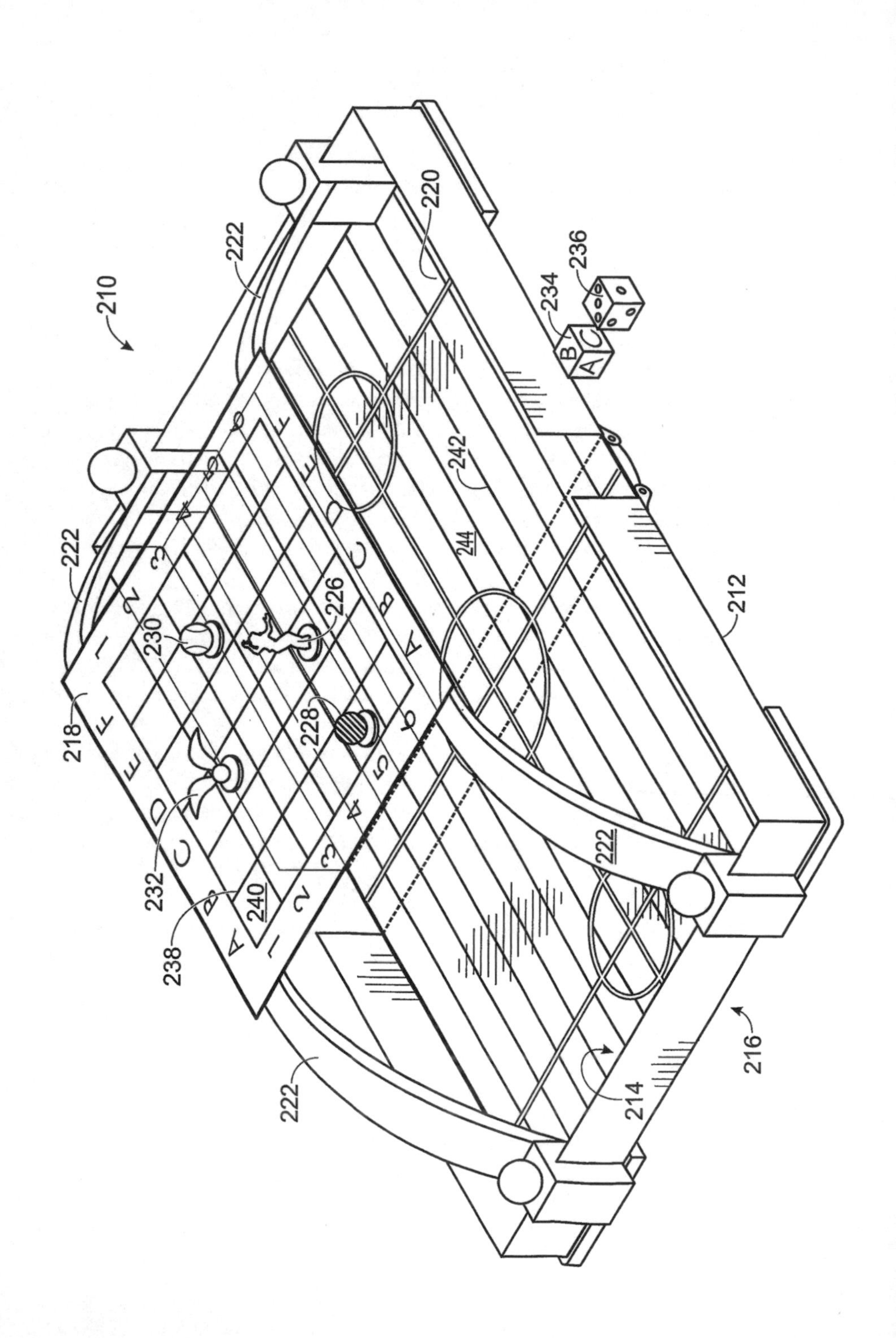
210
212
214
216
218
220
222
226
228
230
232
234
236
238
240
242
244

board game. It owes a debt, perhaps, to Monopoly, the first game involving movement around a board, and which also involved collecting cash on each circuit.

The official American database contains 40 live trade marks containing 'Harry Potter' for various itemised products or services. This is apart from words such as 'Dementor' (cruel creatures in the series) which also turn up as trade marks. What is interesting is that there are a number of 'dead' entries for those that were never registered, or which lost registration. These are for wordings which are not in the canon of books, such as 'Harry Potter and the Battle of Hogwarts'.

The British official trade mark site lists three other as yet unpublished 'titles' that have been filed by Bloomsbury, Rowling's UK publishers: 'Harry Potter and the Chariots of Light', 'Harry Potter and the Pyramids of Furmat', and 'Harry Potter and the Alchemist's Cell'.

My guess is that these were intended as smoke screens, filed so that when the seventh and final book in the series was being written, its title (*Harry Potter and the Deathly Hallows*) would remain secret. It seems odd, though, that the American and British filings were different.

At the time of writing, the Wizarding World of Harry Potter theme park is about to open in Florida, to further exploit the ideas and names used in the books. With no new output from Rowling on the horizon, the licensees may be taking a risk, but Potter's creator is the wealthiest author ever.

Game with multi-level game board, WO 2003/037460, published 8 May 2003

To Arthur! 250 years of the black stuff

In 1759 Arthur Guinness signed a 1,000-year lease on the St James's Gate premises in Dublin. In 2009 successor company Diageo threw a party with the slogan 'To Arthur!' to mark the 250th anniversary. Here are two inventions by the company for pouring its famous black, bitter stout: one famous for its fans, one that few will be familiar with.

Just about everyone has heard of the Guinness 'Widget'. This is a small device incorporated into a can of stout which, on pouring, gives a nice creamy head without any special equipment. It is placed at the bottom of the can and emits a gas. It won the Queen's Award for Technological Achievement in 1991 for an invention which had been worked on since at least 1959. The patent application was published in 1987 as GB 2183592, 'A beverage package and a method of packaging a beverage containing gas in solution'.

It is also well known that Guinness takes a long time to pour. In September 2002 Diageo trialled a dispensing system called Guinness FastPour which claimed to deliver a perfect pint in only 25 seconds instead of the usual two minutes. It was tried out in pubs across London and Yorkshire, but was dropped after opposition from loyal drinkers: they enjoyed the wait.

Then Diageo came up with a new technique which it was promised would give drinkers at home the perfect head on a pint of Guinness. The concept had been tested since 2003 in Japanese bars, which are often too small to accommodate the usual keg-and-tap system for pouring stout. It uses a device which emits ultrasonic pulses through glasses holding specially prepared stout. The device is the Guinness Surger (illustrated) and it went on sale in some UK stores in early 2006.

The kit as sold consisted of a pint glass and two full cans as well as the mains-powered unit itself. A little water is first poured on to the unit's metal plate. Then the contents of a special battery-equipped can of stout (using a different gas mix from that found in 'normal' cans) are poured into a pint glass and placed on top. The device is then switched on and takes between 30 and 90 seconds to create what the company describes as a 'velvet pint with a creamy head'.

Technically speaking, the unit uses an ultrasonic transducer to produce 'excitation' and hence cavitation of the liquid, which encourages the gas in the liquid to come out of solution. The water on the plate is there to ensure that there are no efficiency-reducing air gaps between the bottom of the glass and the plate.

Diageo spent £2.5 million advertising the concept. They were presumably conscious that the move towards non-smoking in public would encourage the existing trend to drink more at home rather than in pubs. Even in bars, where the Surger was also marketed, many younger drinkers prefer premixed drinks, and busy bar staff must find it a nuisance waiting so long for a pint to settle.

The cost of the apparatus – £16.99 – was a potential deterrent, and users had to remember to buy the correct cans. The 30–90-second preparation time was seen as less of a problem, for regular Guinness drinkers are used to waiting, 119.5 seconds (it is officially estimated) for the 'perfect-pint' ritual to be completed – an Irish variant on the Japanese tea ceremony, perhaps.

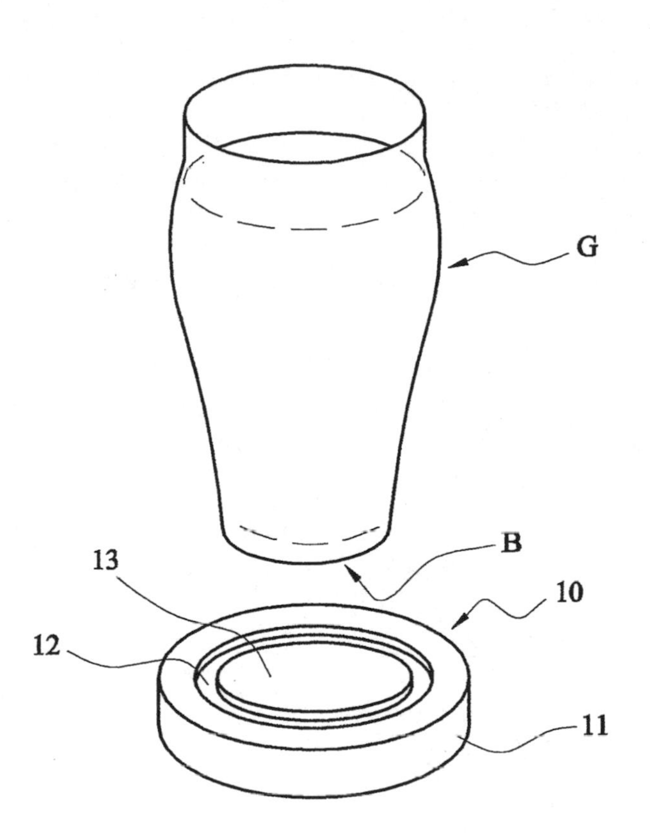

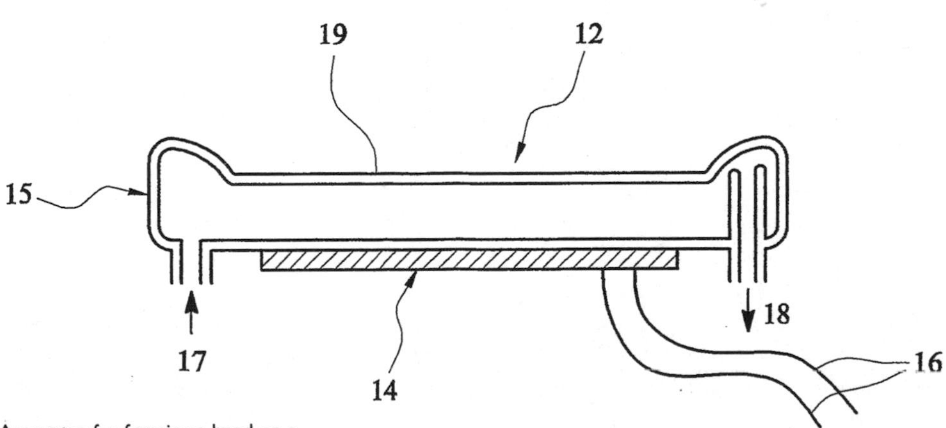

Apparatus for forming a head on a beverage, WO 2004/011362, published 5 February 2004

Indeed, the company's advertising has cleverly made a virtue of the delay, with slogans such as 'Good things come to those who wait' and endless television shots of entranced would-be drinkers staring at the stout slowly settling. Diageo hoped that their at-home variation on the theme would create 'the theatre and anticipation around the Guinness' that many drinkers expect and enjoy.

It didn't. Sales of the kits were lower than anticipated and the home version was withdrawn in Britain in April 2008. The cans are still available for those who own a unit, and a bar-top version is still used in some other countries.

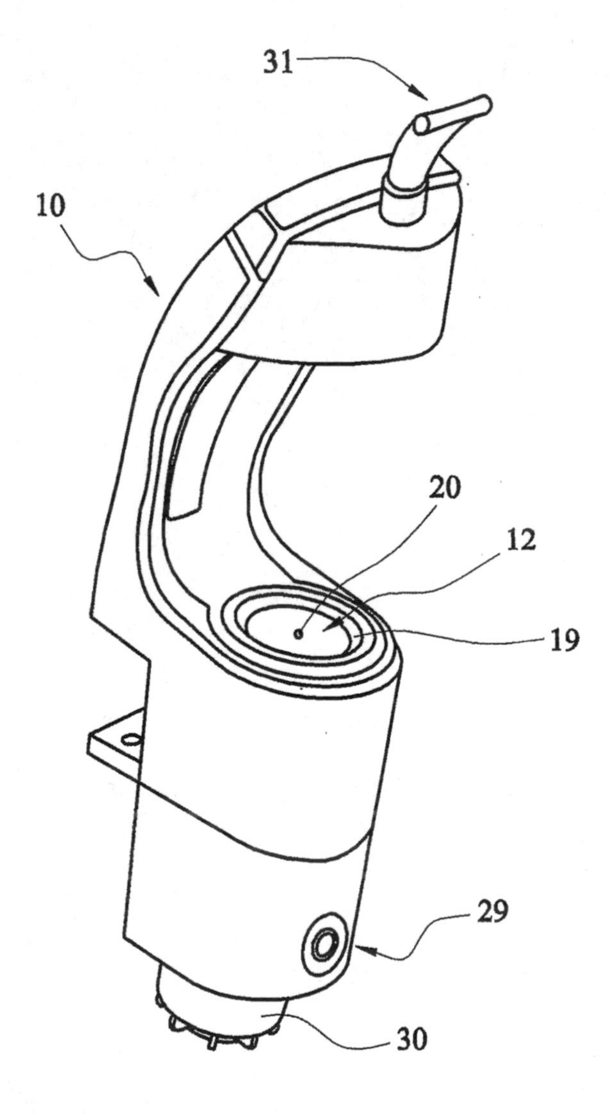

The electronic nose

The tragic bombings in London on 7 July 2005 prompted many to call for bag searching at the ticket barriers on the Underground. To do this would, however, cause huge delays and require a great deal of additional manpower.

A possible alternative is using an 'electronic nose' to sniff out explosives so that only bags deemed suspicious need to be searched. The electronic-nose concept has been around for a long time, and was initially ridiculed. The basic idea is for a device that can sense the specific components of an odour and analyse its chemical make-up to identify it. For example an array of electronic sensors could detect odours and then a second array would attempt to recognise the pattern.

Typically, electronic noses have been large and expensive. Current research is focused on making the devices smaller, cheaper, and more sensitive. The smallest version, a 'nose-on-a-chip', comprises a single computer chip containing both the sensors and the processing components. Applications are currently in use in the food, beverage and cosmetics industries, among others, and research is being undertaken with a view to detecting specific diseases such as TB and lung cancer.

It is likely that patents will arrive focusing on precisely this idea, just as the sinking of the *Titanic* in 1912 provoked numerous patents for methods of detecting icebergs, evacuating ships and the like, and the 9/11 attacks spawned patent applications such as WO 03/026960, which explains the inventor's ideas about separating the flight deck of an aircraft from the passenger cabins so that access between the two is impossible during flight – all explained in Japanese, incidentally, though it seems a simple enough idea.

One of the leading inventions in the field is the zNose by Electronic Sensor Technology, based in California. According to the company, 'The Model 4500 zNose is the only mobile, small footprint analyser on the market that can detect and canalize accurately organic, biological & chemical compounds in real time.'

The company claims that its zNose technology is a breakthrough in the area, and has demonstrated it to members of the US Congressional Committee on Homeland Security. Apparently only 4% of cargo containers entering the USA are routinely monitored in this way (it was 2% before 9/11). The company states that unlike conventional trace detection systems, which detect only a small number of specific chemical elements, the zNose is a 'true electronic nose' designed to recognise the unique chemical fingerprint or olfactory signature of any odour or fragrance. It can be trained adaptively to recognise the chemical profile of materials commonly used by terrorists.

The invention consists of a gas chromatograph using surface acoustic waves (SAW). A silicon collector (shown in the illustration in an 'exploded' view) concentrates the gases and feeds them to a capillary column together with an inert gas such as helium for separation and identification. As the chemicals travel along the capillary they interact with the column's chemical coating and are slowed by the interaction. Since each interaction is different, each chemical component exits the column at a different time. The column effluent is focused into the SAW resonator for chemical detection. A visual display enables even minimally trained operators to determine whether further action is needed. It all sounds rather like magic to the uninitiated.

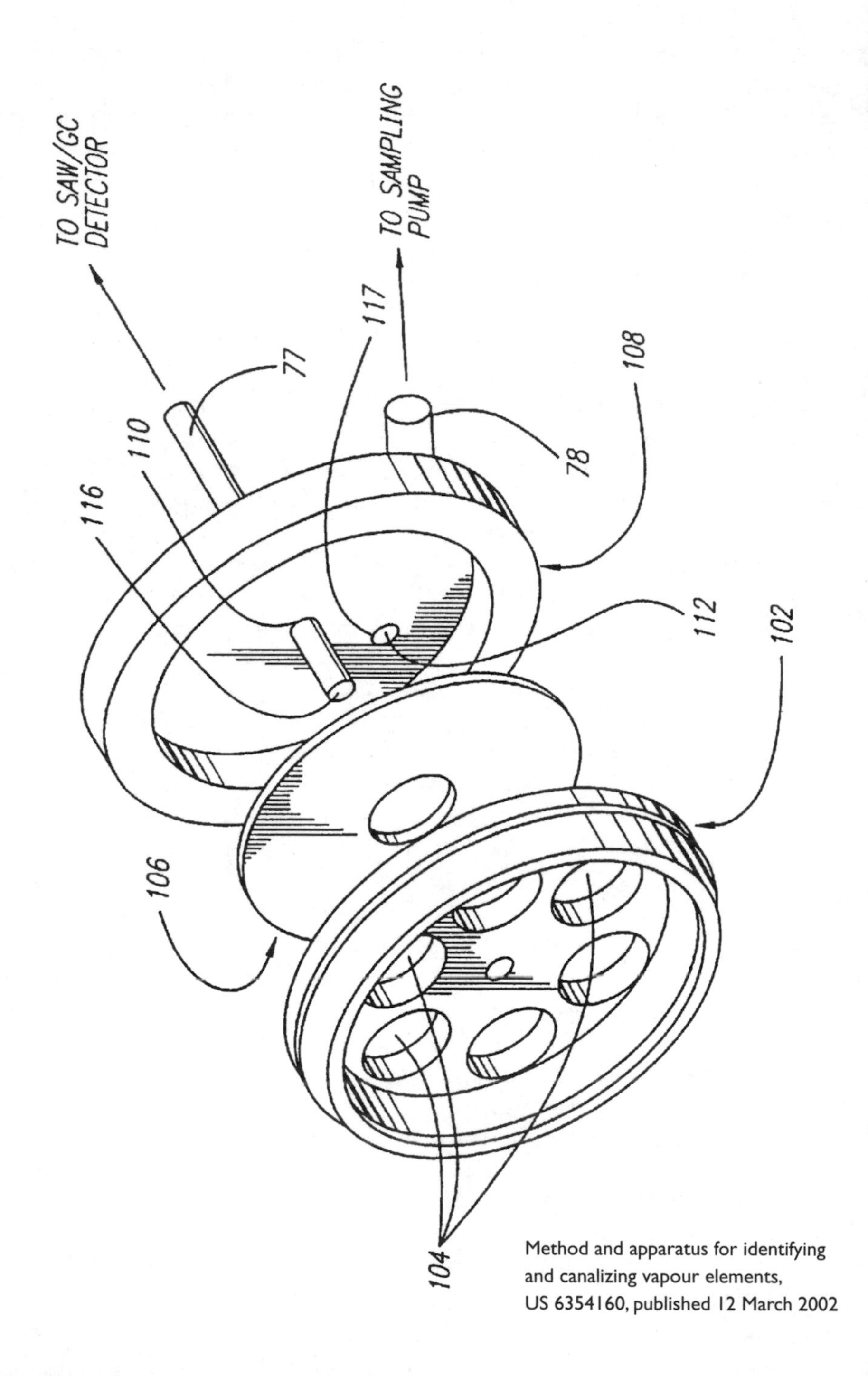

Method and apparatus for identifying and canalizing vapour elements, US 6354160, published 12 March 2002

The device takes 10 seconds to gather sufficient information about an odour. This would make it ineffective for monitoring passing passengers in busy locations such as stations and airports, but it could prove useful in analysing static targets such as cargo containers. The major components of the device are shown below.

Dr Edward Staples, the inventor and the managing director of Electronic Sensor Technology, has been applying for patents in the area since 1991. If the product is as versatile as he believes then the number of possible uses must indeed be huge. The company claims that the technology is covered by four American patents, and that it has already been used by law-enforcement and military organisations across the world.

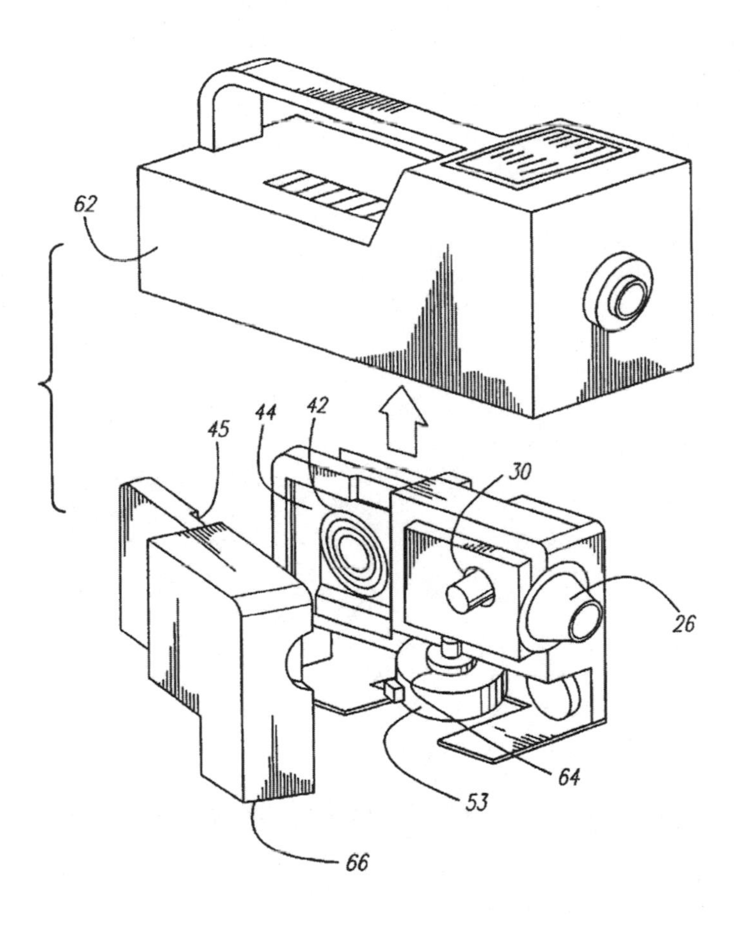

Product placement

The idea of featuring branded products in films and on television has been around for a long time.

An early example is when Mars was offered the opportunity to pay for one of their products to be featured in the 1982 film *E.T. the Extra Terrestrial* – the scene where Elliott leaves a trail of sweets to tempt the unknown creature out of the shed. Mars considered E.T. to be so ugly that they refused to allow M&M's to be used in the film, believing the creature would frighten children. This allowed Hershey the opportunity to supply their Reese's Pieces. They paid a million dollars, even though they were not allowed to see the script. Sales rose by 65% after the film came out.

A particularly blatant example of product placement is the 2006 James Bond film *Casino Royale*, which stars the latest incarnation of 007, Daniel Craig. A girl asks him pointedly about the watch he is wearing, for example, and Sony's products are especially prominent, as they had gained the distribution rights. Some estimates suggest that over £45 million was received in cash (or in kind) for product placement in this film alone.

Product placement, or 'embedded marketing', had in a sense already been present (but unpaid-for) in Ian Fleming's James Bond books. Fleming was forever pointing out that his hero smoked Dunhill cigarettes, drank Beefeater gin and wore Rolex watches, supposedly to show that Bond lived in the real (if posh) world.

Out of curiosity, in November 2009 I visited the official database of so-called 'world' patent applications, Patentscope, and searched the full texts for 'James Bond'. Seventy-five hits were returned, many of which by their titles sound very relevant to the secret agent ('Electronic mail-based adventure game and method of operation', 'Flying all-terrain vehicle'). One looked quite intriguing: 'Automatic generation of trailers containing product placements', published in 2006 by electronics company Philips.

This is a deliciously cynical invention. It explains that television is a great way to promote products, but that 'many people see the commercials as a break for making sandwiches or going to the bathroom'. To get round this, marketing executives arranged for product placement within the actual shows. There is now greatly increased choice with the variety of channels and Internet viewing, so any film or programme is seen by fewer and fewer people, hence 'reduced eyeballs for the product placement providers'.

The patent specification suggests that trailers could be used to circumvent this problem, using automatic generators on 'home media servers'. Such generators would use editing techniques to ensure that the product placement content remained present. The product placement data segments would be tagged within the 'superset' of data in the production stage to ensure that it appeared in altered versions. This would provide 'a better service to the providers of the product placement data'.

Further, all this could be adapted to the culture, language and so on of a particular region, or even to the gender, age and habits of a particular viewer. (This sounds very ambitious: how would they gather that information? The method is not explained, so far as I can tell, in the specification.) The advertiser for the product could make a bid for a particular kind of customer, and could specify whether all segments should be included or only a certain percentage.

A counter would keep track of which product placements were chosen and how many times a product placement was chosen and viewed. Based on the value of the counter an invoice would be created, with usage information being automatically passed on whenever the user logged on to the Internet. 'The providers of the product placements will be ensured [*sic*] that, even if the viewer decides not to see the film, the viewer would have watched their product in the trailer that was generated'. That is a great relief to all of us, I am sure.

British television may soon have much more product placement. In late 2009 a consultation period was announced by the government during which the possibility of allowing independent commercial television to use product placement in their programmes would be considered. This would put an end to the curious anomaly that, for example, the popular commercial-TV soap opera *Coronation Street* must feature in its breakfast scenes an imaginary cereal (Nutty Flakes) and sell beer supplied by an imaginary brewer (Newton & Ridley) in the Rovers Return. Commercial television is suffering from a fall in advertising revenues and it hopes that this measure will help. Children's programmes, however, would remain subject to the product-placement ban.

The proposal has many opponents. Simon Hoggart of the *Guardian*, for instance, has described the concept as 'a form of corruption, by which elements of our favourite shows are covertly sold off to the highest bidder without our being told'. The Labour government agreed to permit the concept provided (after opposition from within the Cabinet) that exceptions were made for alcohol, junk food and gambling. The attitude of the new coalition government is similar.

Inventions in the Universal Studios theme parks

I grew up in America, but have only visited one theme park there – Universal Studios, in Los Angeles, in 1992.

I recall a tram ride, which took us through a subway tunnel set. We were told that luckily the unit using the set was upstate filming when – oh no, an earthquake! It rumbled, floods gushed in and a bus crashed down towards us. Great fun, if bad for those with weak hearts.

Here are a few of over 30 patent specifications by Universal Studios. Some of their inventions are for routine applications, such as the 'Guest positioning assembly' which locks the passengers into cars during a ride. Others are more exciting, and can be linked to films made by the company. A big problem for the attractions is that the effects must be repeatedly provided for each group of visitors, and so must be easily reset.

Their 'A flame barrier, apparatus and method for entertaining guests' is a modification of the popular Revenge of the Mummy ride. The patent points out that routinely fire effects are either simulated, if close to the visitors, or project the flames away from them, 'thus limiting the guest's real sense of danger'. At the time the flame effects were 7 to 10 metres away. The invention was for a transparent barrier protecting the visitors in their cars so that the flames could safely be brought closer to them. Electrostatic means for removing the resultant soot were also provided. Other cleaning methods, such as using a squeegee, were mentioned.

The Men in Black Alien Attack ride is based on the technology set out in the first-illustrated patent. The basic idea is that visitors are provided with weapons from which to fire light beams at the aliens and so accumulate points. The passengers in each car compete with those in the other cars, and can hinder them by exceeding point totals. A computer in a control room is linked to the targets and also to track sensors which control the movement of the cars. None of this is related to the plot of the film *Men in Black*, of course.

I am not sufficiently familiar with Universal's films to be able to determine from which film, if any, the second-illustrated invention comes. Contrary to appearances, this is for a ride during which the passengers do not get wet. The patent document explains that some rides divert waterfalls to the side, and refers to Disneyland's Big Thunder Mountain Railroad and their Jungle Cruise. In both cases the car veers away from the waterfall at the last moment. In this invention, the opposite happens: the waterfall is diverted away from the car just before it passes through and then resumes once it is past.

Lastly, *Twister* has clearly inspired the 'Amusement attraction with man-made tornado', where the occupants of a house experience a cyclone. Again it has to be capable of being recreated for the next group of visitors, since after 'the tornado experience has ended . . . the guests proceed to the exit'.

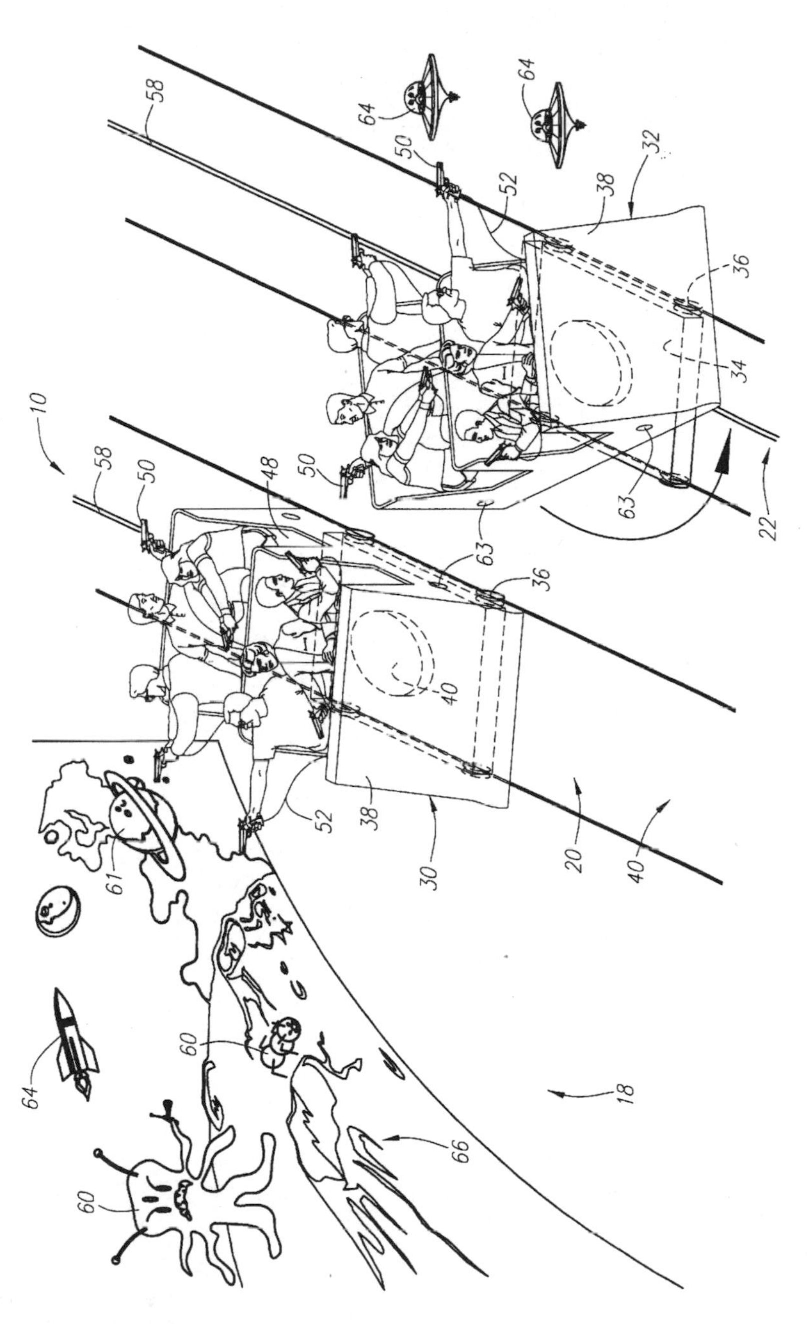

Amusement system, US 6220965,
published 24 April 2001

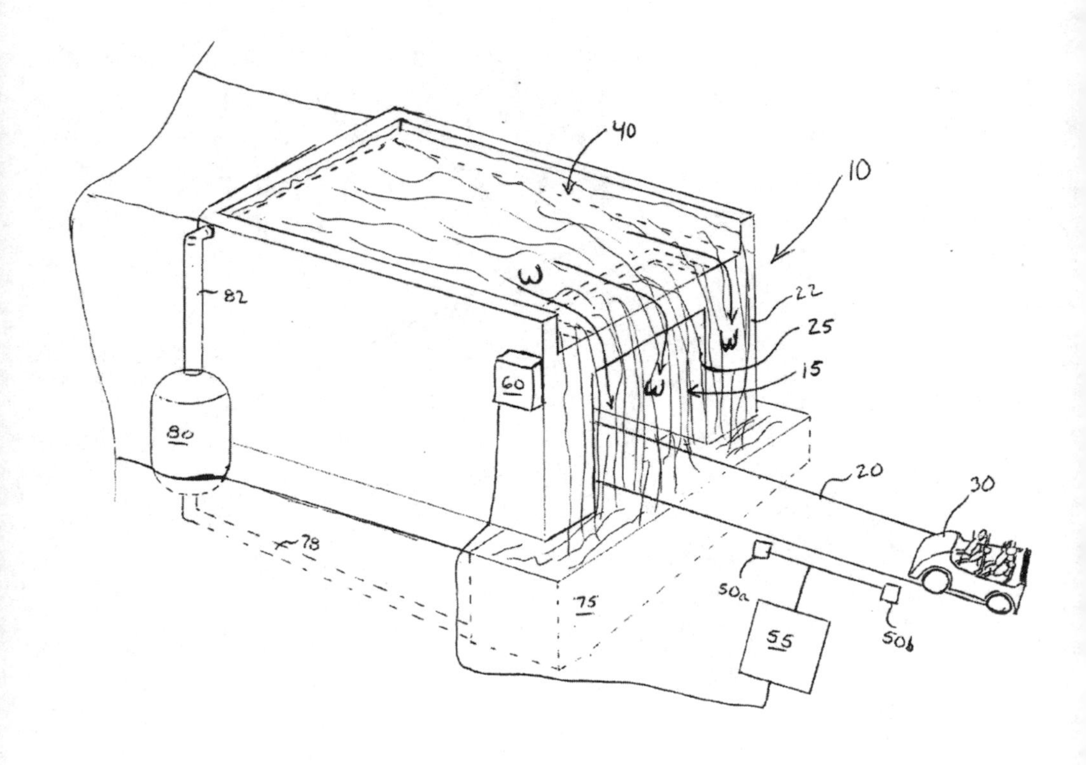

High-speed, punch-through waterfall effect, US 2007/093305, published 26 April 2007

Amusement attraction with man-made tornado, US 6254489, published 3 July 2001

Inventions on *Dragons' Den*

This reality television show originated in Japan (as *Money Tigers*) in 2001 and the format has since been sold to many countries. The British version began in 2005 and has been very popular as an example of actual, get-down-in-the-dirt-and-fight business activity: negotiating finance for a product or service. Reality TV with an entrepreneurial edge.

The supplicants are forced to walk up a staircase to face the Dragons in a bare, warehouse-like interior, presumably to render them even more breathless when they make their pitch. There are five Dragons, wealthy entrepreneurs with cash piled up next to their chairs, who listen to a two-minute presentation which they are not allowed to interrupt. They then ask questions and decide whether they want to invest all or part of a sum requested by the applicant in return for a share of the equity. 'I'm out' is the response if a Dragon decides an idea does not merit investment. Only if all the money requested is offered can the applicants receive it. Often more important than the money are the skills and contacts of the Dragons, as many aspirants realise.

If the idea fails to make money the Dragons lose their investment, which some applicants do not seem to appreciate, convinced that they are providing a favour. Dragon Theo Paphitis is known for pointing out that his children's inheritance is at stake. The show is not just about inventions, but of course these are the ones that interest me. Many are sensible ideas, if not always well presented, but it is the less likely notions that are often most interesting.

The Dragons rightly enquire about intellectual property protecting the product, but have been known to ask if a world patent exists, to which the confident reply is usually affirmative. Actually the term 'world patent' is meaningless, signifying merely a blanket application for rights which are then dealt with separately in each country. Trade marks rarely get the attention they deserve in the show as valuable selling tools.

When speaking with inventors I often refer to the programme as providing an example of the kind of hard questioning that backers will subject them to. Why should anyone risk their money for a poorly worked-out idea, presented without facts and figures? Inventors need to think about the motives of the person sitting opposite them. It is, of course, a disadvantage if the inventor is a poor communicator, or very nervous, and *Dragons' Den* can cruelly expose any unfortunate sweating his or her way through a hesitant 'What am I doing here?' speech. The television cameras and the hot lights merely add to the strain of confronting five staring faces.

One inventor on the show was Roark McMaster, of Hayling Island, with his Q-Top cucumber shield, intended to prevent the sliced end from drying out. There was much hilarity among the panel, as they clearly thought that McMaster had come up with a solution to a non-existent problem. 'What about cling film?' suggested one. Nevertheless, Peter Jones offered him a job because he admired the inventor's skills.

The main drawing from the patent application, which was terminated before grant, is illustrated. Made from moulded EVA foam, it is shaped like one end of a cucumber. The product is presently on sale as a 'cucumber lemon cap bottle stopper'. For a time McMaster's website promised an 'end to oxidation misery'.

In a similar if less versatile vein is the Banana Guard, which was pitched for on a Canadian version of the show. It is a banana-shaped,

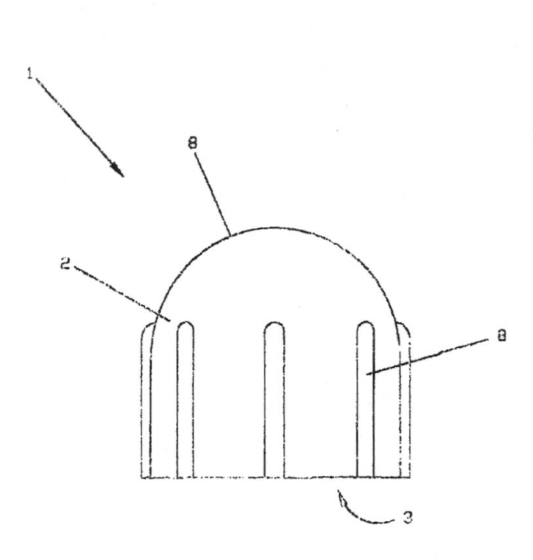

Cap for covering a cut end of a food item, GB 2425297, published 25 October 2006

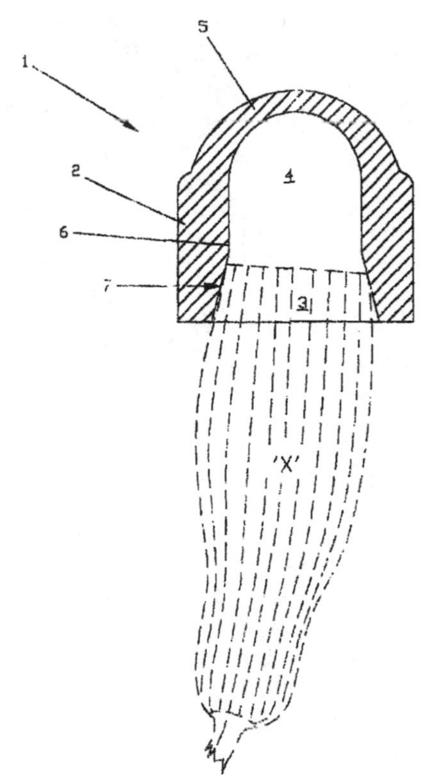

moulded plastic container that snaps shut to keep the contents fresh. An offer from the Canadian Dragons was declined, though production went ahead. The company website says it was designed to 'prevent banana trauma'. Appropriately enough, with that wording, it was designed by doctors working in a Vancouver casualty department. It has sold a million so far and the company has since expanded into Froot Case and Froot Guard, which are for spherical fruits, and plans a similar item to keep sandwiches fresh.

Samantha Gore turned up twice on the British show, with crime prevention techniques (fake television in an empty house, then automatic curtain drawing in the same), neither of which worked when demonstrated before the dragons. These inventions did not get an investment. She has three published patents specifications so far. Another inventor's egg cooker failed to impress when the inventor forgot to put the egg in.

Duncan Bannatyne, one of the Dragons on the British show, tells of a pitch made by a lady inventor, Gayle Blanchflower of London, for cardboard beach furniture. He asked what would happen if it rained. He was told by the inventor (with a 'look of hatred') that he was stupid – people don't go to the beach in the rain. He persisted. What happened when a child came out of the sea and sat in one of the chairs? The reply was to keep the child under

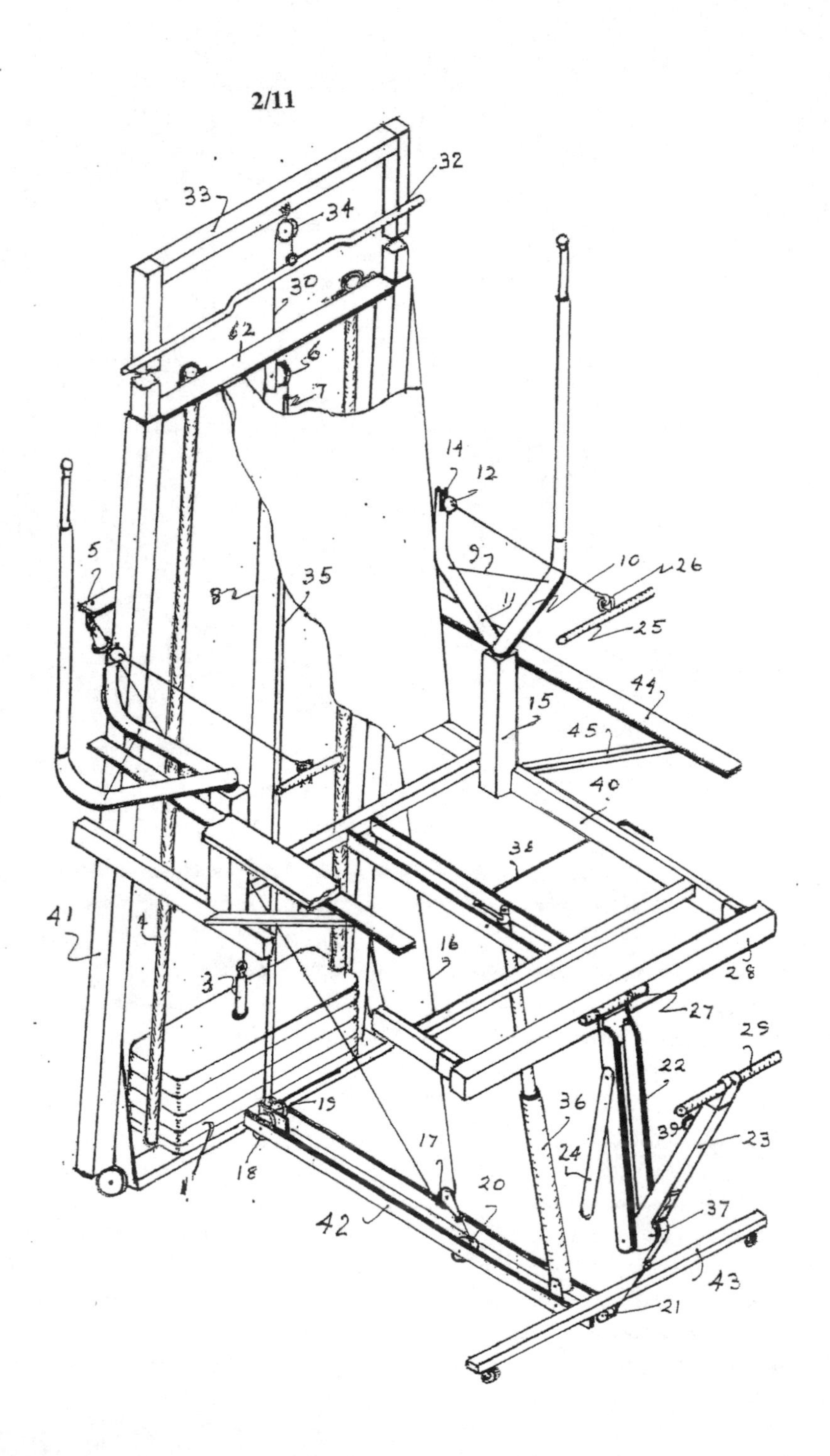
2/11
33
34
32
30
62
6
7
14
12
9
10
26
11
25
5
35
82
15
44
45
40
38
41
4
16
28
3
27
29
22
19
36
39
23
1
18
17
24
20
37
42
43
21

control. As so often, the inventor appeared to have assumed that the product would be used only in perfect conditions.

Blanchflower had spent £60,000 on intellectual property rights. Her idea was published as 'Portable article of furniture', in which lightweight items collapse down to fit in a bag.

Finally there is the illustrated Easy x chair by Peter Ashley of Felixstowe. It is a gymnasium concealed within a lounge chair. It reduced the normally poker-faced Bannatyne to tears when he tried to use it. Ashley is now refining his new office-chair multigym, which is equipped with hydraulics. 'This one,' he says, 'will be a winner.'

We shall see.

Chair type exercise apparatus, WO 2004/067107, published 12 August 2004

A novel broom

Many small children wonder why something is done in a certain way, and innocently suggest a better way. It is not often that something comes of the suggestion.

Samuel Houghton of Buxton, Derbyshire was just three back in 2006 when he was watching his father using a broom in the garden. 'He was doing the leaves and I asked him why he was using different brushes and then I went into the garage and invented. Then I called Daddy to go into the shed and then I said, "Why swap the brushes when you can use this?" and that was my invention.'

The invention consists of two brushes arranged in a scissor manner. Two brooms mean that the debris is more likely to be picked up – if the first broom misses something the second one may catch it. Preferably, the first broom picks up coarse material, the second dust, as the bristles are different in size.

Samuel's father Mark just happened to be a patent attorney. These are professionals who write patent specifications and argue their merits if the patent office suggests the idea is not new. Based on his client's instructions, Mark wrote the description of the invention and based on that wrote the claims to what was new that included language such as this first, main claim:

> A sweeping device for sweeping a surface, the device comprising a combination of two brushes connected by a resilient connector; the first brush comprises a brush head, a plurality of bristles affixed to the brush head and a handle extending from the brush head; the second brush comprises a brush head, a plurality of bristles affixed to the brush head and a handle extending from the brush head, wherein the resilient connector serves to retain said combination of brushes in resiliently moveable relation to one another in use.

The language is written in 'patentese' to make it as precise yet as broad as can be allowed. Using language that was too specific could allow someone else to 'design round' the patent.

Unusually, the precise language of the description ends with 'Acknowledgement of the inspiration to invent (but in no means to the present invention per se) and the ability to identify an invention, and announce it as such, is hereby given to Archie the Inventor of Balamory.' Archie is an inventor who lives in a castle in the popular live-action *Balamory* children's TV show. Samuel likes Archie because 'Balamory is a real place, and Wallace and Gromit are just made-up inventors'.

The published application includes a search report which lists anything similar. 'People said it was done before and Daddy said, "See for yourself," and they couldn't find anybody,' says Samuel. Indeed, the search report does not contain any citations to similar patents. The broom has since been granted protection in the UK.

All this might seem like a stunt to promote Mark's work as a patent attorney. Neither he nor his wife, Susan, a special-needs teacher, feels that way. Mark works for a firm of attorneys which only represents business clients. He admits that inventors rarely see their patents become manufactured products. 'Patenting is a negative right,' says Mark. 'It stops other people from doing things.' However, even for a company 'to have a head start on their competitors' can itself be valuable.

Since his first invention, Samuel has found that a child's trumpet makes a useful plug for the bath as he can pull it out himself, unlike the existing plug. 'It's not an invention,' says Samuel. 'It was just using something for a different thing.' Actually, doing so can indeed lead to a valid patent.

In a *Guardian* interview father Mark was asked if they would like to find a manufacturer to get the broom into the market. 'It would be nice if we did, but we are doing this to help Samuel learn, help him understand a little bit about innovation,' says Mark.

'No,' corrects Samuel. 'I did it to help Daddy.'

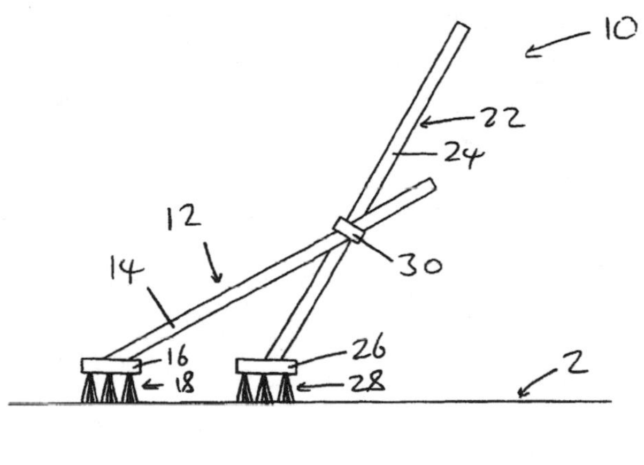

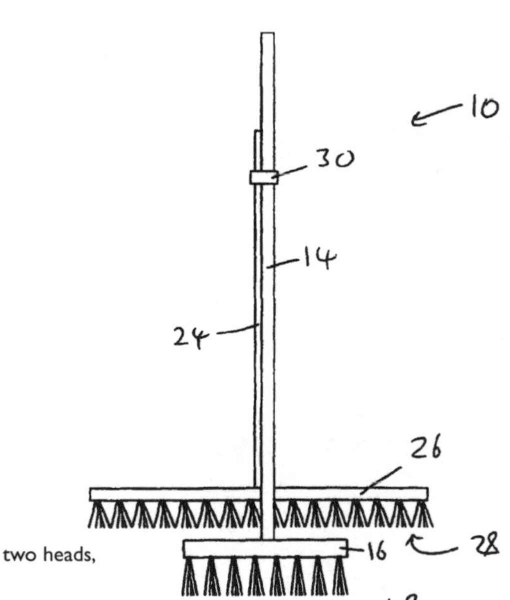

A sweeping device with two heads, GB 2438091, published 14 November 2007

The yoomi self-warming baby bottle

Following the birth of their first, premature child, who needed small quantities of milk around the clock, Jim Shaikh and his wife Farah found night-time feeds challenging.

He would go off to prepare the milk but had trouble getting the temperature right. Frequently he would overheat the milk and would then have to run the cold tap over the bottle. While cooling the milk, he would sometimes nod off, only to wake and find the milk was too cool. Farah meantime was waiting upstairs with the crying baby, wondering why he was taking so long.

'You're an engineer, can't you fix this?' implored Farah. A few days later Farah remarked that it would be wonderful if there was a bottle that heated itself to the right temperature. Her husband agreed, and proceeded to work out the basic principles on a pub napkin.

Shaikh's experience as an engineer, first at BMW Rover and then as managing director of his own consulting firm, Intelligent Fluid Solutions, gave him the tools to experiment with the idea. Before developing the first prototype, he tested the viability of the concept while earning his MBA at London Business School. Realising that yoomi met a need in the market, he began developing the product in earnest. He also started the patent application process.

After five years of hard work and help from family, friends and a slew of experts, aided by grants from the London Development Agency, the product was born. The Shaikhs' second son, Niall, became the bottle's first test subject. The yoomi passed, and in 2008 was awarded the London Technology Fund's Design and Engineering prize.

The bottle contains a warmer that, after charging, is activated by pressing a button. This triggers the internal solution to warm the feed in just 60 seconds. The cold feed is gently warmed as it flows through specially designed channels on the outside of the warmer. By the time it reaches the teat, it is at natural breast milk temperature (32–34°C). Warming the feed in this manner to only breast-milk temperature avoids impairing the nutrients by overheating or heating too rapidly.

The bottle uses a 'phase change' material (a material that changes from a liquid to a solid) which generates heat as it solidifies. This material, commonly used in hand warmers, is a non-toxic, concentrated salt solution (it is used as an additive in ice cream). As an additional safety precaution, the internal solution also contains both a food-grade colourant so that any leaks are visible, and Bitrex, a non-toxic but bitter substance that will ensure that, if there is a leak, the baby will refuse the contaminated solution.

The warmer can be recharged up to 100 times by placing either in a pan of boiling water or in an electronic steam steriliser.

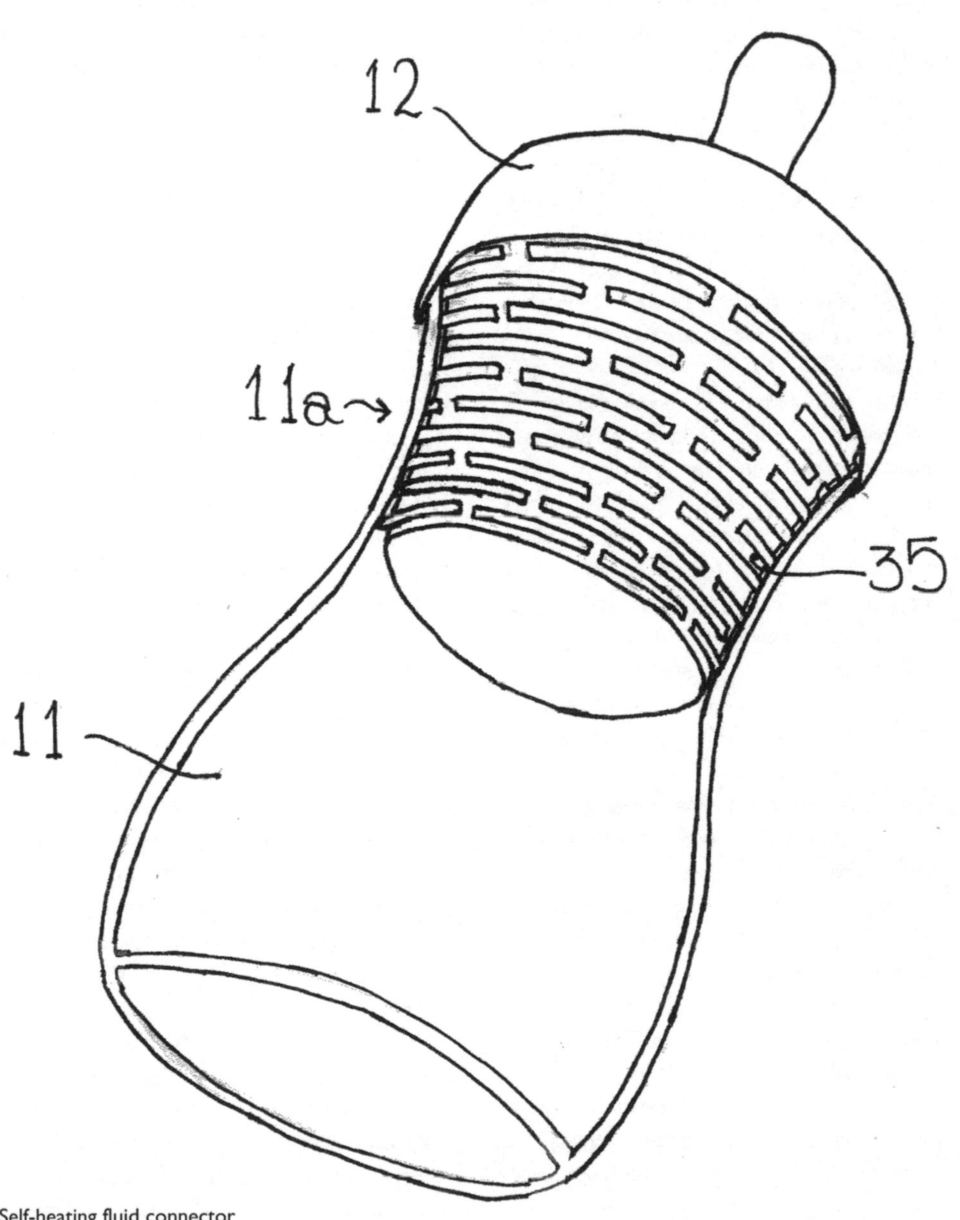

Self-heating fluid connector
and self-heating fluid container,
WO 2006/109098,
published 19 October 2006

Anti-skimming devices

In 2008 there was a story in the British press about a man who used 49 cloned credit cards to steal £10,000 from an automatic teller machine (ATM).

It was late at night, and it took him two hours to do it. Whenever anyone wanted to use the machine, he politely stood back. And then went on stealing money with another card.

What he didn't know was that a CCTV camera was filming him at Barclays Chislehurst branch, so police were hoping to catch him. I do not know if they did.

Over £535 million is lost annually in Britain due to credit-card fraud. Both the card details and the PIN are needed by criminals. Much crime at cash machines originates from 'shoulder surfing', where criminals watch numbers being inserted, and either pick-pocket the card afterwards or take the card after distracting the victim by, say, dropping a banknote.

Another method is to use card-trapping devices. A plastic loop is inserted into the card slot which traps and retains the customer's card. The 'Lebanese loop' is so-called because it was allegedly first used by a Lebanese gang. The criminal poses as a helpful bystander and suggests that the victim re-enters the pin, which he memorises. After the card-holder gives up and leaves, the fraudster removes the device, along with the card.

Card details are being purloined by increasingly sophisticated means. This could be a 'skimming' device placed over the card slot, which captures card details from the magnetic strip. In addition, a miniature camera hidden inside a strip of metal fixed unobtrusively to the top of the machine watches as the customer punches in the PIN number. Details from both are sent by wireless communication to criminals who typically will be using a laptop in a van parked near by. Sometimes an entire simulated frontage complete with dummy ATM is erected. Once the details have been gathered, a cloned card is swiftly manufactured and put to bad use.

The police say that many of those targeting ATMs in this way are criminal gangs who typically buy legally available pinhole cameras and card-reading equipment of the sort used in some workplace controlled-entry systems, and then adapt them for their own ends.

The banks, of course, are not just sitting idly by while all this is happening. The British banking industry has, since 2002, been funding a specialist police unit, the Dedicated Cheque and Plastic Crime Unit. Banking industry fraud investigators work side by side with the police.

In November 2004 Barclays announced that they had begun trials of an anti-skimming device that had virtually eradicated ATM fraud for banks using it on the Continent, and they claim to be the only UK bank to have installed such devices.

Details are understandably confidential, but a search on the Internet suggests that it is a physical barrier which prevents spying equipment from being installed, or which closes down the machine if an attempt is made to do so. In January 2008 there was a report of anti-skimming devices being smashed at a Newbury branch of the bank, which put the machines out of order.

In 1967 the world's first ATM (other than a 1939 experiment in New York) was installed in a north London branch of Barclays. It required the use of an envelope containing a special cheque embedded with a radioactive compound of carbon-14. This was matched against the PIN entered on a keypad, and cash was then paid out. A proposed PIN length of six digits was rejected in favour of four because this

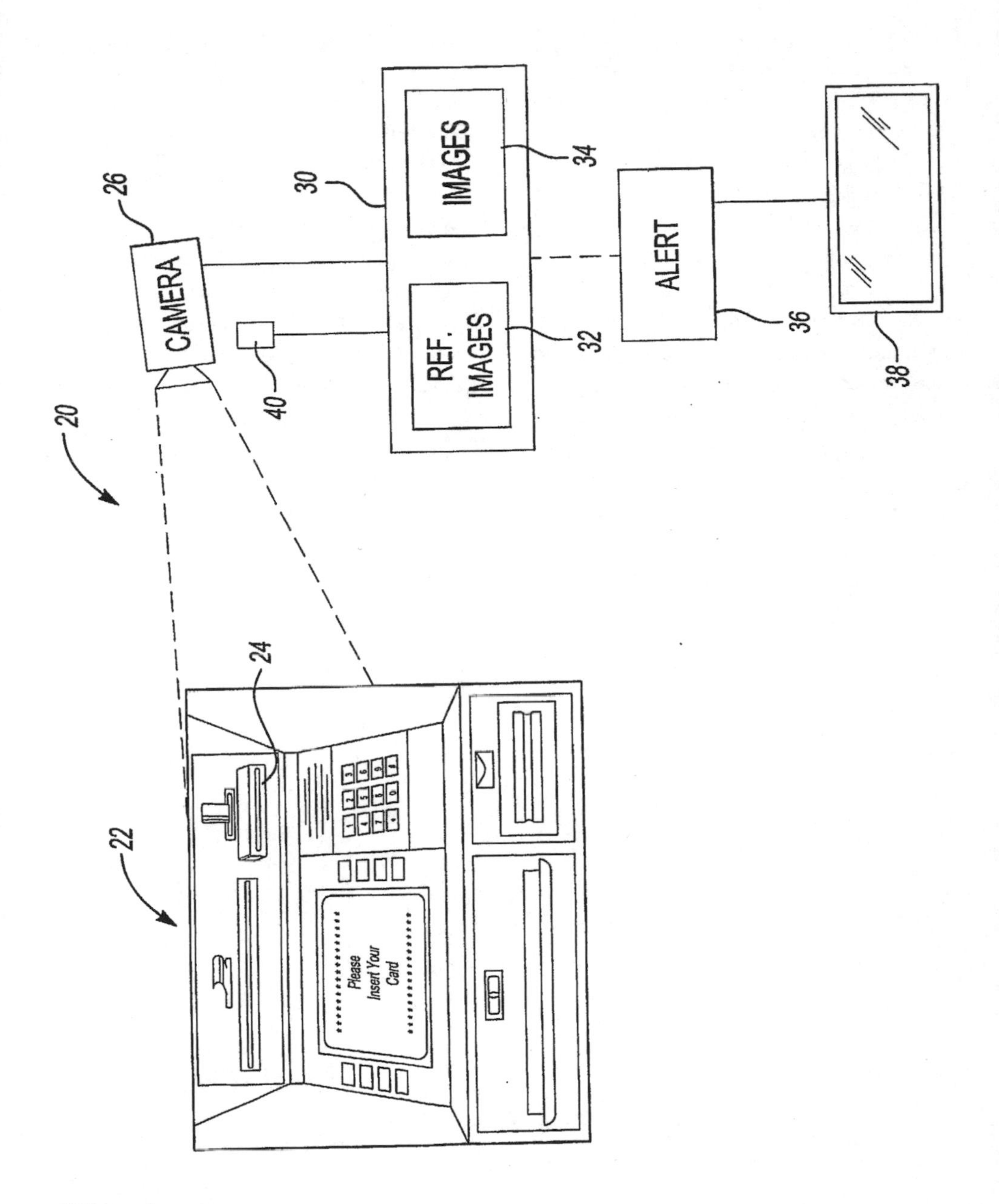

ATM security system,
WO 2005/109315, published
17 November 2005

was the longest string of numbers that the inventor's wife could remember. The concept was not patented for fear of criminals studying the patent details.

There are some patents for anti-skimming equipment. British patent application 2421300 by a private inventor detects a reduction in light if a dummy front is placed over it.

More sophisticated is the application shown in the illustration. A camera stores images of what portions of the ATM should look like. At intervals fresh images are taken and compared with the reference image. If the image is different then the ATM is closed down and an alert is transmitted. Different lighting conditions or the time of day are taken into account in the reference images.

The applicant is UTC Fire & Safety Corporation, a Connecticut company. The increased costs arising from the installation of such devices are, inevitably, passed on to the consumer.

Other methods make use of biometrics, where stored data about the customer is compared at the ATM. Fingerprints are favoured for the majority of inventions, though some use other aspects such as retinal scans or voice recognition. US 6424249 is an authentication system using all three which has been heavily cited as similar by patent office examiners in later patents. In June 2007, Barclays Bank and the NCR Corporation jointly announced that they were introducing the first biometric-enabled ATM in the United Arab Emirates. The struggle between criminals and counter-measures continues.

The fight over high-definition recordings

An important factor in the drive for more capacity in storage media is the attraction of high-definition (HD) video. The problem has been that HD recordings require much more storage capacity than conventional media.

There have been continual efforts to squeeze more and more on to storage discs. On average, an audio CD requires about 8.8MB per minute, a DVD video 35MB per minute, and a HD video over 200MB per minute. This means that a standard 4.7GB DVD will only be able to hold about 24 minutes of HD video.

Any attempts to enable HD technology are likely to involve 'disruptive technology' – the scrapping of old equipment and assembly lines. Quite apart from the problem of encouraging customers to buy new equipment, there is the dilemma of which format they should buy. Blu-ray disc and HD DVD were the leading contenders. We have been here before, in the struggle in the late 1970s and the 1980s between video formats Betamax and VHS. Betamax was technically superior but was unable to record longer programmes such as films, and was generally outmanoeuvred in the publicity war.

In February 2002 Sony, Matsushita and Philips announced their new DVD concept, Blu-ray. I believe it is that patent application illustrated here, which was applied for by all three companies. Oddly, it carries a 'priority date' of the initial filing for May 2002, three months after the announcement. Perhaps they were still finalising the details when they made it.

Four Japanese and two Dutchmen are listed as the inventors. It uses a 0.1mm disc substrate layer that allows up to 23GB of storage on one side of a DVD. That requires new tooling and equipment, raising production costs.

The disc player is backwards compatible with (i.e. can play) DVDs or CDs. It uses violet-blue lasers which have a shorter wavelength (450 nanometres) than the red 650-nanometre lasers used in conventional DVD players, so that substantially more data can be stored on a Blu-ray disc than on a DVD.

Because Blu-ray places the data recording layer close to the surface of the disc, early discs were susceptible to contamination and scratches, and had to be encased in unattractive plastic caddies for protection. This would have been likely to hurt sales, and the discs were subsequently made with a layer of protective material on the surface through which the data could be read.

The first titles were released on Blu-ray in June 2006. The format was exclusively supported by Columbia Pictures and MGM, both owned by Sony, as well as by Disney, 20th Century Fox and Lionsgate. Toshiba's rival HD DVD was exclusively supported by Universal Studios and the Weinstein Company. Others were willing to release material in either format.

What was, perhaps, a fatal blow in the struggle for market share was struck in June 2007. Blockbuster, Inc., the largest chain of video game and DVD rental shops in the world, announced that its next batch of high-definition recordings would be available only in the Blu-ray format. Dell and Sony (of course) hammered another nail into the coffin when they announced that their products would only support the Blu-ray format.

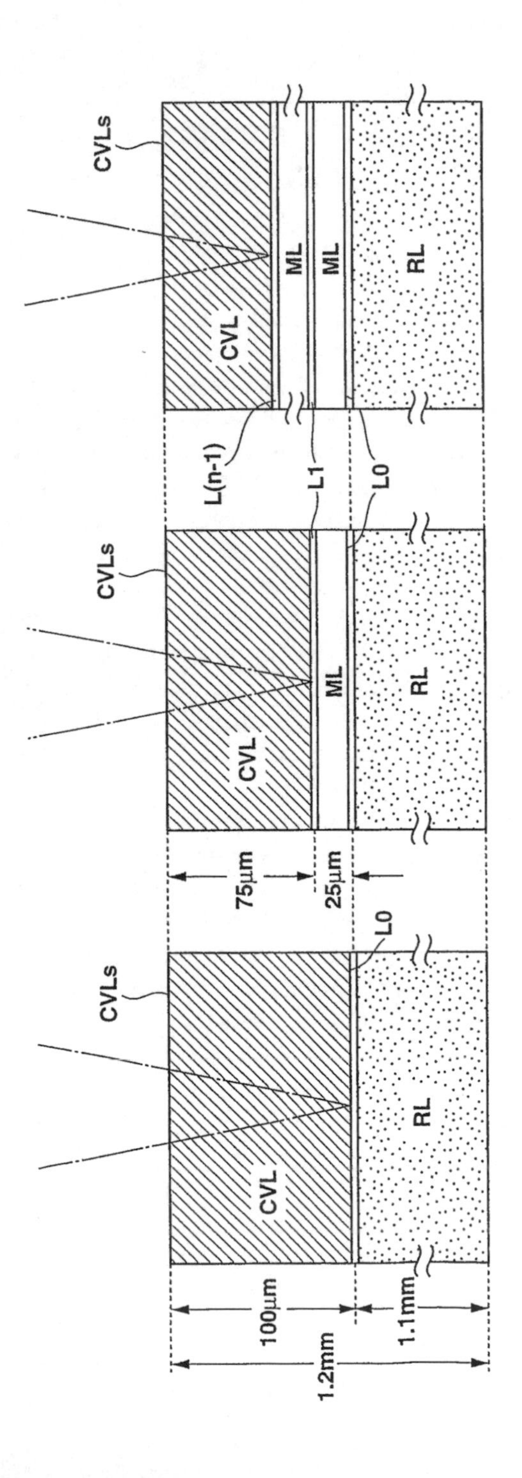

	LAYER ADDRESS
FIRST LAYER : L0	0
SECOND LAYER : L1	1
nTH LAYER : L(n-1)	n-1

In February 2008 Toshiba gave up the uneven battle. At the time Blu-ray discs were outselling HD DVDs by three to one. 'It was an agonising decision for me, but I thought if we kept running this business it would have grave ramifications for the management of our company,' said Toshiba president Atsutoshi Nishida. 'We made a quick decision, judging that there is no way of winning the competition.'

Sales for Blu-ray have not been as spectacular as was hoped. European sales of Blu-ray discs in the first quarter of 2010 were 8.4 million, as against 135 million DVDs. There is the problem of persuading consumers to ditch their old players for one that admittedly has more capacity but where, some argue, the picture is not all that much better, which is especially hard in a deep recession. There are also those who are waiting for another scenario to unfold: HD that arrives at your computer via the Web, to be viewed on its monitor or on 'HD-enabled' television.

Bill Gates commented back in 2005 that Blu-ray would be 'the last physical format there will ever be', as soon 'everything is going to be streamed directly or on a hard disk'. Of course, Gates hasn't always been right – he once predicted that e-mail would lead to paperless offices.

Disc-shaped recording medium, disc driving device and disc producing method, WO 2003/100702, published 4 December 2003

The Segway personal transporter

Dean Kamen of New Hampshire-based DEKA Products, a self-taught engineer, was already wealthy from earlier inventions such as portable dialysis machines and drug-infusion pumps when he asked his company to work on the Ginger project. It was a spin-off from their six-wheeled, stair-climbing wheelchair.

Hints crept out about the project, and Kamen proclaimed that Ginger would be the main mode of transport in a decade. Such luminaries as Steve Jobs of Apple and Jeff Bezos of Amazon clamoured to be backers. Jobs said he thought it would be as big a deal as the PC. The engineers working on it were just as obsessed.

It was finally unveiled on 3 December 2001 on ABC's *Good Morning America*, and so began the pilgrimage to New England by police forces and the US Postal Service, among many others. A 2002 episode of *Frasier* shows Niles Crane deftly manoeuvring back and forth on Kamen's invention, which must have been a great publicity coup.

Others, though, were disappointed after so much hype. 'It won't beam you to Mars or turn lead into gold,' Kamen said. 'So sue me.'

The object of all this attention was the two-wheeled Segway scooter (Kamen hates that word), which has a tiny platform to stand on and its movement controlled by ten microprocessors, several gyroscopes, two batteries and a host of software.

The rider moves by leaning in the desired direction while gripping a handlebar, and stops by standing upright. First, though, 'speed limiting is accomplished by pitching the vehicle back in the direction opposite from the current direction of travel, which causes the vehicle to slow down'. This is what the patent says: clearly the user must be alert at all times to prevent unwanted motion. The batteries are rechargeable overnight, and the top speed is electronically limited to 20 kilometres per hour (a version produced for the police is capable of twice that speed).

Kamen envisages cars being banned from city centres in favour of 'empowered pedestrians'. He is poetic in describing what the Segway can do. 'When you use a Segway, there's a gyroscope that acts like your inner ear, a computer that acts like your brain, motors that act like your muscles, wheels that act like your feet. Suddenly, you feel like you have on a pair of magic sneakers, and instead of falling forward, you go sailing across the room.'

Sales, however, have not been good. By the middle of 2009 they had reached only 50,000 worldwide.

This is partly due to the problem of where the things can be legally used. The project was well advanced before anyone thought about the problem of whether state transport laws permitted their use on pavements. Is it a road vehicle (in which case riders are highly vulnerable, and a menace to other road-users), or should it operate on pavements (where it could endanger slower-moving pedestrians)? For a time the US Occupational Safety and Health Administration wanted to classify the Segway as a 'powered industrial truck', but many American states have passed legislation permitting their use on pavements.

In Britain, a strict interpretation of the 1835 Highway Act means that what the Department of Transport calls 'self balancing scooters' are not allowed on pavements, while EU vehicle

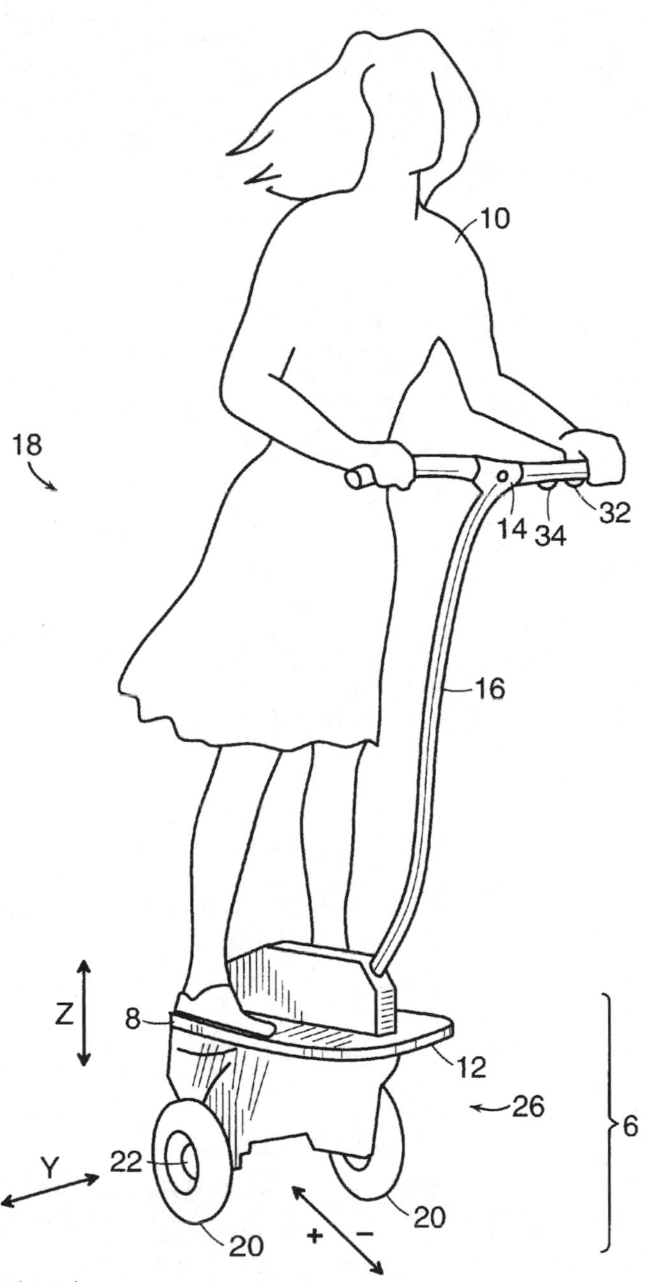

Personal mobility vehicles and methods, US 6367817, published 9 April 2002

certification rules mean that they cannot be used on roads. A price tag of about £5,000 has not helped either. It has done better on the Continent, where its use on pavements is widely allowed. There are models which can carry golf bags, and off-road versions for more rugged terrain. They are also in use commercially, enabling staff to move quickly around large premises such as warehouses and airports.

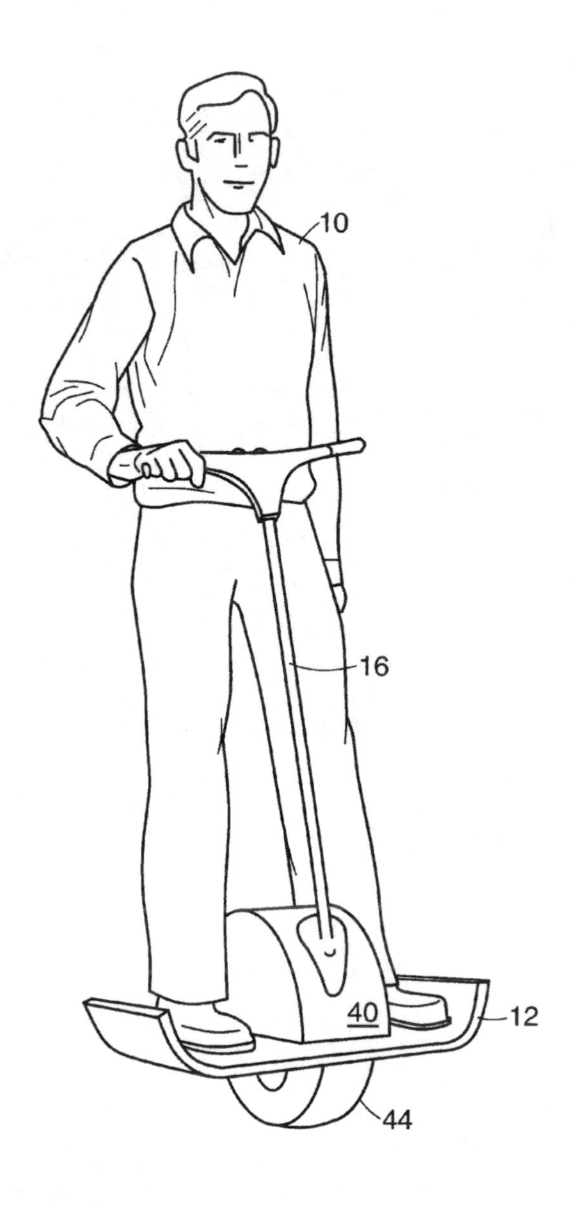

Self-cleaning glass

The science-fiction writer Arthur C. Clarke once said that any sufficiently advanced technology is indistinguishable from magic. To me, this can be said of self-cleaning glass. Pilkington Activ glass has an extremely thin coating of microcrystalline titanium oxide which absorbs ultraviolet light from daylight. This causes a photocatalytic reaction, which breaks down dirt on the glass with no need for any cleaning agents. When water falls on it, a hydrophilic ('water loving') effect is created whereby the droplets attract each other, so that the dirt slides off within a sheet of water. The glass dries cleanly – an effect which the inventors like to call the 'invisible squeegee'. Condensation is also greatly reduced.

'Pilkington Activ is based on titanium dioxide, which is used in foodstuffs, toothpastes, and sun cream,' explained Dr Kevin Sanderson, one of the four inventors who developed the glass at Pilkington's technical research centre in Lathom, Lancashire. 'But usually it is a white powder, which is not ideal for glass because you can't see though it. So we used it in a thin film form – 15 nanometres thick – so that it appears as close to normal glass as it can.'

Although not quite nanotechnology, the coating (and the chemical reactions) can be regarded as being at the nano-scale, just one thousand millionth of a metre.

As no chemicals are needed to clean it there is an environmental advantage, and although it may threaten window cleaners' jobs, at least they will no longer have to risk injury working at high level, nor cause disruption by their presence, although in fact some cleaning is still required by conventional means.

Marketing efforts target in particular buildings where large expanses of glass are difficult to reach using normal methods. It has also been used instead of reversible windows – the type which tilt in either direction to aid cleaning. The glass has also been found useful for car wing mirrors, where dirt and condensation are greatly reduced.

The innovation is a result of a long process of research and development over the previous decade into 'thin film' technologies. 'When we realised we could get these properties with it, that is when it really drove through,' said Dr Sanderson.

It is more expensive than conventional glass, adding about 15 to 20% to the cost of installation. A study by Britain's Building Research Establishment of 12 buildings, published in 2006, concluded that it was cost effective when lifetime expenditure on cleaning and so forth was taken into account. Both project managers and the occupants were happy with it – including the zoo that is now cleaning its glass monthly instead of daily.

The Roomba robot vacuum cleaner

With over three million units sold since its 2002 debut, the Roomba vacuum cleaner has been a big hit. It is the first popular usage of robots to assist around the house.

The idea of a robot vacuum cleaner dates back to the 1970s at least. The illustration is from the original patent and is by Massachusetts company iRobot, which was founded in 1990 by three former MIT staff who had perhaps been inspired by Isaac Asimov's 1950 novel *I, Robot*. The company has published over a hundred patent applications in the field of controlling the movement of 'robotic vehicles'. Based on their work, iRobot has made models capable of cleaning roof gutters or swimming pools, and also a bomb-disposal model, the PackBot. This is currently deployed in Iraq and Afghanistan.

The illustration shows the ideas behind the first-generation model, with buttons shown in the first image for S, M and L (small, medium and large rooms). Component (23) is the bumper. The second illustration shows the underside, with (34)–(36) being the power system, (78) the bristle of the side brush, and (88) the cleaning head. The control module, motive power and sensors are not shown.

The sensors are, of course, key. In this first model their role was to identify objects encountered by the bumpers rather than to detect the size of the room. The bumpers would recoil on contact; this would activate the sensor which would notify the control module. The 'bounce' mode would then initiate – in other words, it would move back. The rechargeable power pack could provide up to 90 minutes of power. The lack of a power cord was certainly an advance on ordinary vacuum cleaners – the patent points out that the problems of coping with an electric cord would 'severely degrade' performance.

The company has been constantly improving the product, and that model was superseded in 2004. The small 'dust cartridge' in the earlier patent application was replaced, better software enabled the calculation of room sizes, there was faster charging up, and 'dirt detection' was added (so that less time is wasted cleaning already clean areas).

The second- and later third-generation models have a self-charging homebase, shaped to fit the circular Roomba, which the machine automatically seeks out using an infrared beacon. It takes about three hours to recharge. The machines can also be programmed to clean at scheduled times.

Newer models no longer have to crash into an object to detect it; instead they use infrared signals and will slow down when approaching an obstacle. They can also estimate the size and shape of the room they are in from the time taken for a signal to bounce back from the walls, and so determine the best cleaning pattern for the area. Sensors on the underside of the unit prevent it from falling down stairs. They constantly send out signals, and a delay in the returned signal suggests that an abyss has been reached. It then sensibly retreats.

Special sensors are included so that it can follow the line of a wall and turn as required. Its makers claim that it can move over carpet fringes and tassels, though some users have disagreed. An accessory is available that sends a signal between two sensors – placed across a doorway, for example – to keep Roomba within

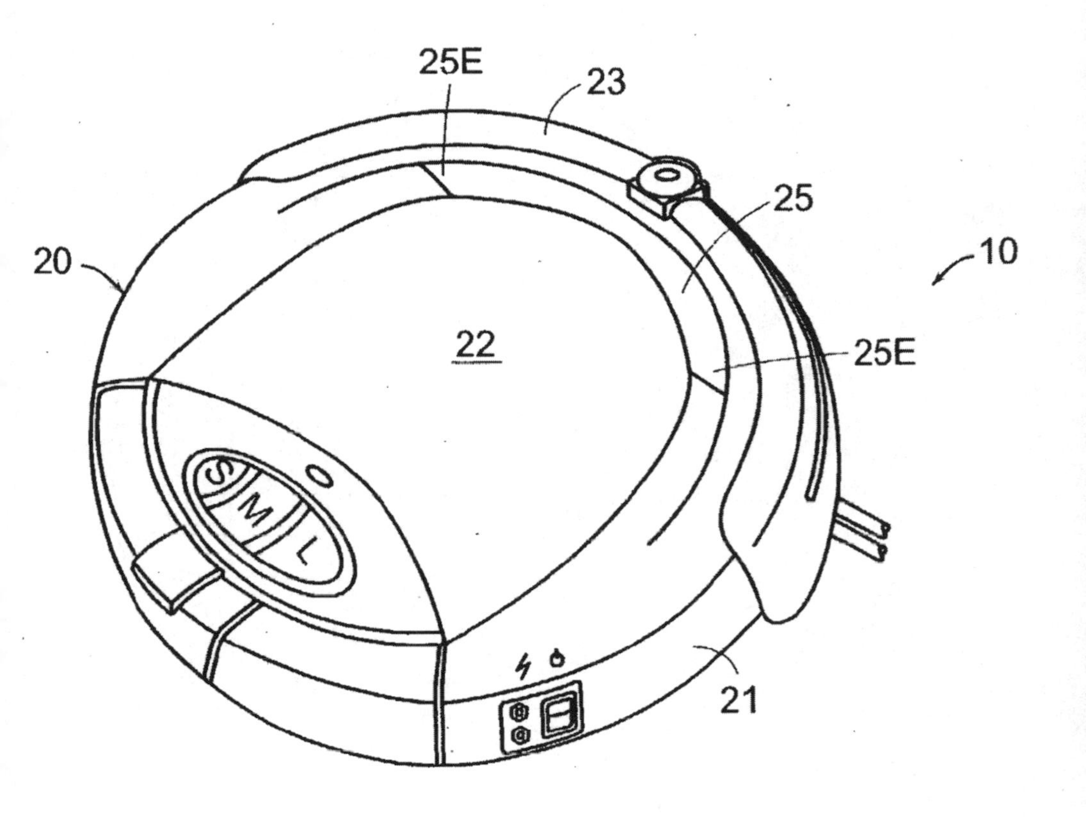

Autonomous floor-cleaning robot,
US 6883201, published 18 March 2004

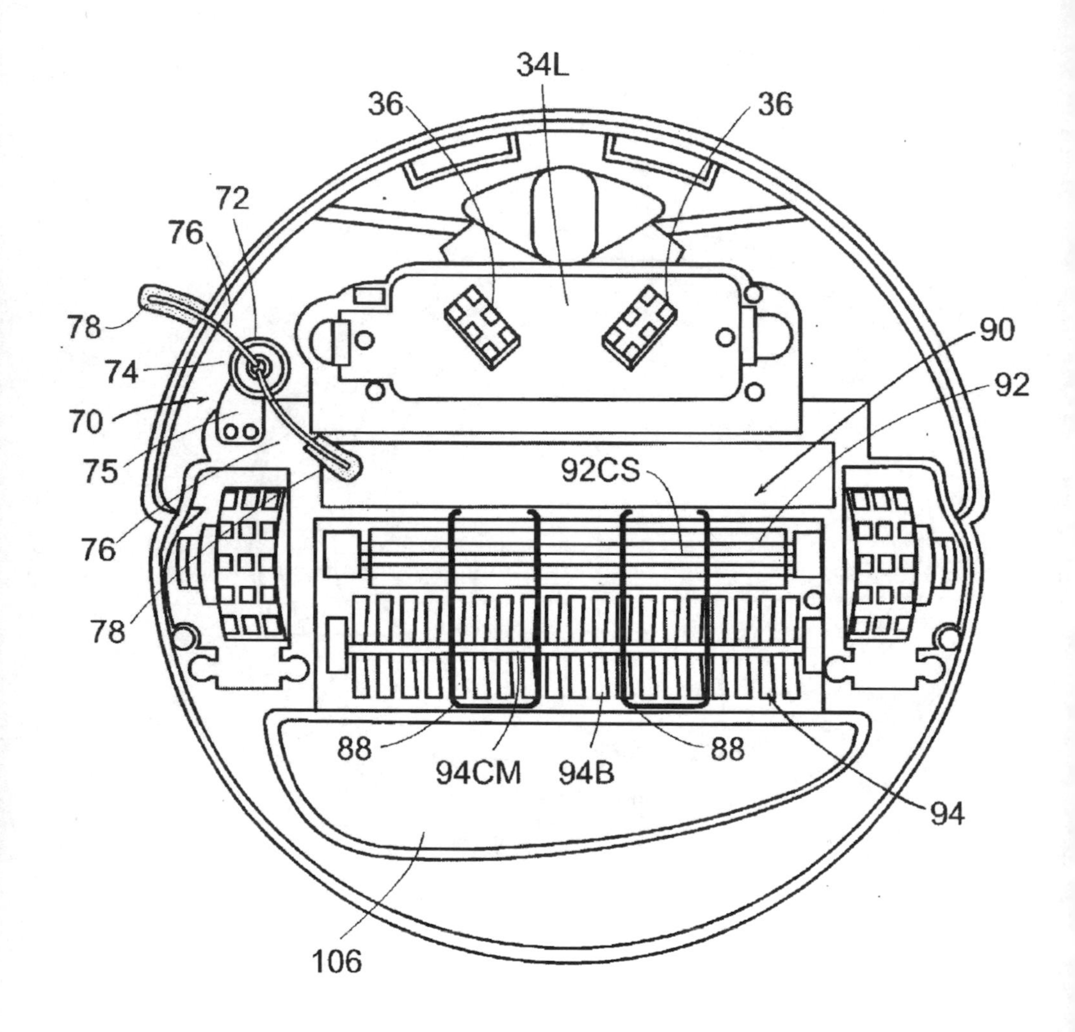
34L
36
36
72
76
78
74
70
75
76
78
90
92
92CS
88
94CM
94B
88
94
106

set boundaries. A new modular design is said to make cleaning, repair and replacement much easier.

Since the machine is designed for unattended operation, it is important that it does not become clogged during use, and it can to some extent declog itself.

Unforeseen problems can occur. One owner returned home to find that her 'pet' had accidentally locked itself in the bathroom after bumping into the half-open door while inside. Another owner says, 'Our cats are curious about it, and one of them just sit[s] on top of it during the recharging.'

So far, you still have to empty the dirt bin. Wet cleaning is provided by the Scooba model which washes, scrubs and dries. As yet, the problems of cleaning anything above floor level, or stairs, remain unsolved.

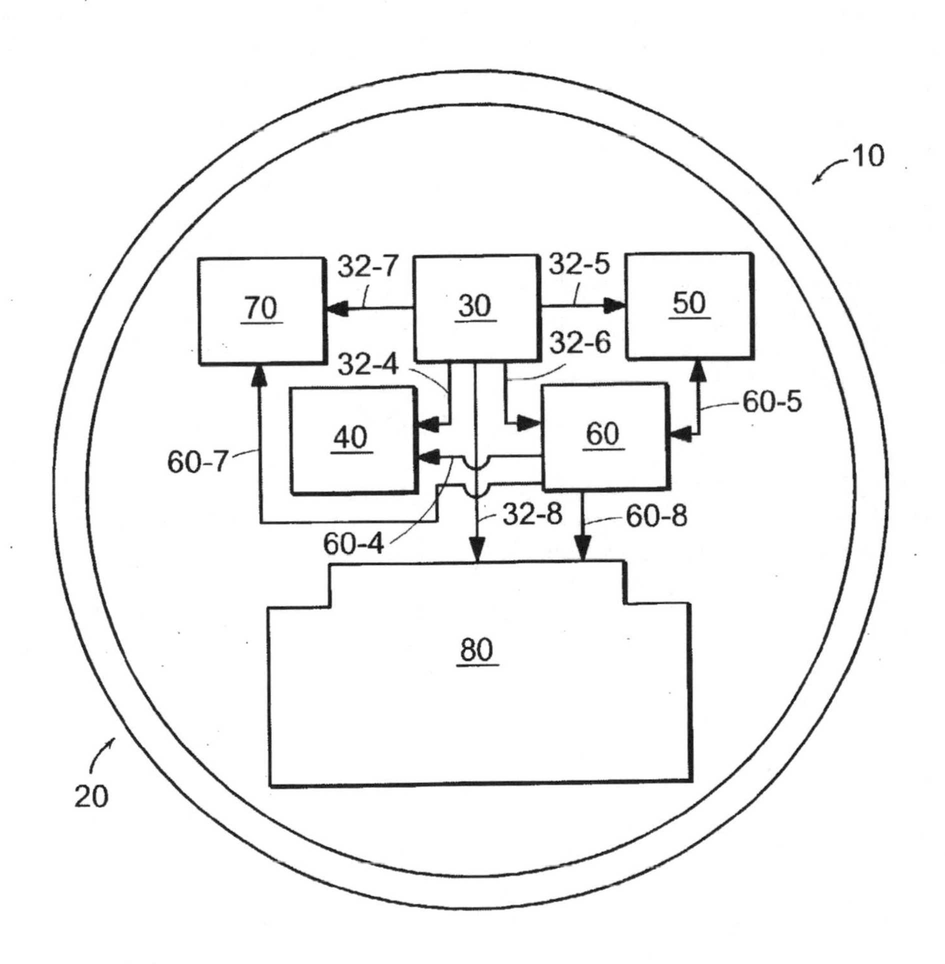

Rotating buildings

David Fisher, an architect based in Milan, has applied for a patent for his 'Rotatable building structure'. It will be constructed floor by floor in a factory, and a floor can be installed on site in six days. Plans are well advanced for the prototype: the Da Vinci Tower in Dubai. According to Fisher's website, there will be 80 floors and it will be 420 metres high. The lower floors will be offices, then a luxury hotel, with apartments at the top. It will require only 600 people in the factory and 80 on-site technicians to build it, instead of the usual 2,000 workers on a traditional construction site of the same size. Only the core of the building will be built on-site. Attractive rippling effects, including the use of lights at night, will be created.

There have been rotating buildings before, such as observation towers and the occasional house. In Fisher's building, however, each floor can revolve independently. The idea was inspired by a remark made by an apartment owner who said that he had the best view in his building. Fisher thought that was unfair.

A mixture of horizontal wind turbines and solar panels will provide the energy needed to turn the floors, the movement of which can be controlled by the occupant – perhaps to take in the views at sunset, or to avoid the midday sun. Excess power would go into the local electricity grid.

The patent document makes interesting reading. For example, each apartment would have two sets of plumbing and wiring connections with the central core. This would mean that one set of flexible connections would stretch, and then temporarily disconnect, so that one set was always available.

The rippling light effects will only be possible if all the tenants agree to move their floors according to a preset sequence. Perhaps their contracts will say that on certain days they must cede control of their floor to the management. In fact, unless the entire floor is owned by a single tenant furious disputes could erupt as one resident's beautiful sunset disappears so that their neighbour may enjoy the view.

If money and sunshine are prerequisites then Dubai was the obvious place to start (though Moscow also wanted one), and Fisher's blueprint looked set to become a reality in the Emirate, with a slightly smaller version planned for the Russian capital. All that was holding up the project was the clarification of land rights. The estimated cost was $700 million and the building was originally scheduled for completion in 2011 by Rotating Tower Dubai Development Limited.

However, with the severe financial crisis hitting Dubai hard at the end of 2009, it is now possible that construction will be delayed, or even abandoned.

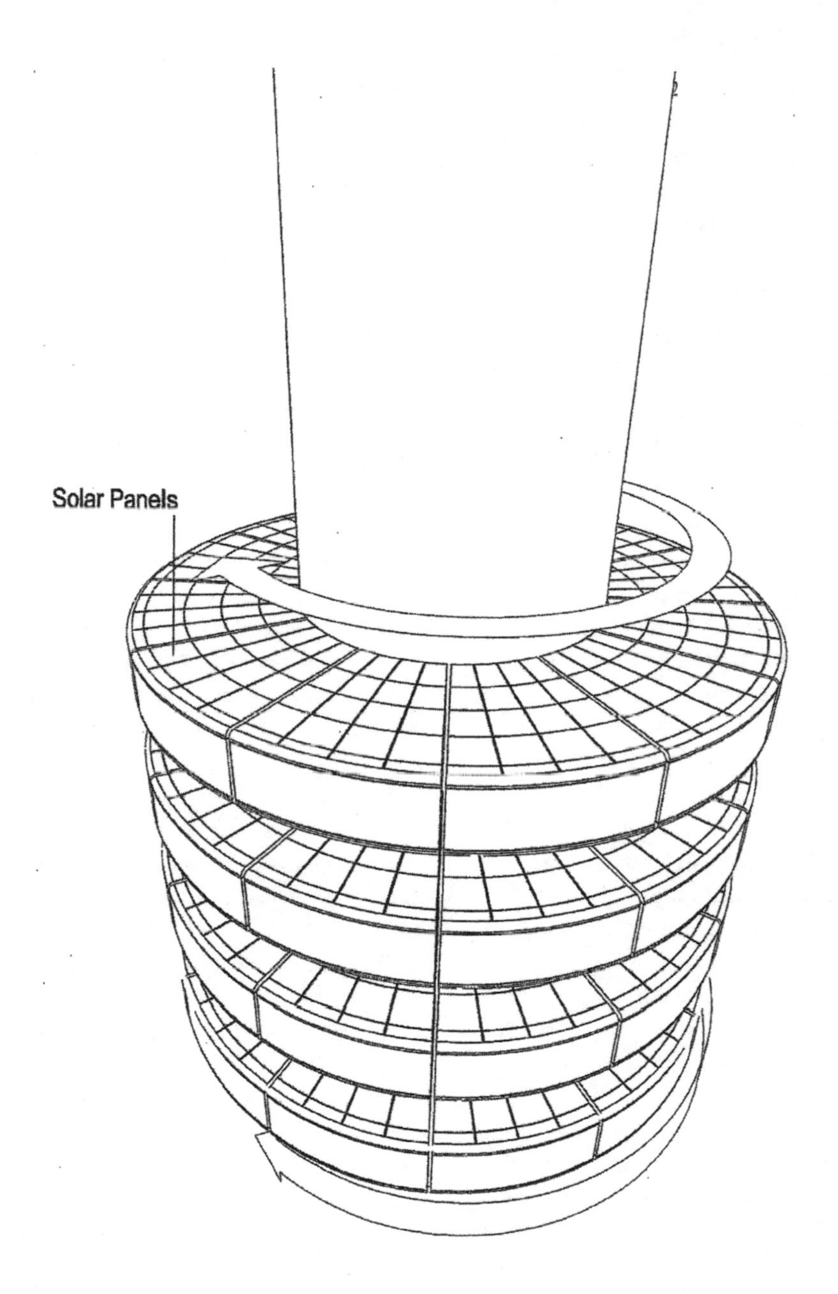

Rotatable building structure,
WO 2007/148192, published
27 December 2007

Automatic number plate recognition

The use by law-enforcement agencies of automatic number plate recognition (ANPR) technology has been around for a while. It was the publicity about its use to identify a car on the night of 30 June 2007 on the M6 in Cheshire, following a terrorist attack at Glasgow Airport, that brought it to prominence.

The concept is to link the heavy use of surveillance cameras in the UK – which, it has been estimated, is home to 1% of the world's population and 20% of the world's CCTV cameras – and passing data from imaged car number plates with a database where 'wanted' plate numbers are flagged up. Nearby police officers can then be alerted to follow the car and arrest the occupants.

There were certainly earlier precedents. For example in November 2005 WPC Sharon Beshenivsky was shot and killed during a robbery in Bradford. The CCTV network in the town was able to use ANPR technology to identify the getaway car and track its movements, leading to the arrest of six suspects.

The idea may sound simple, but there are problems in capturing and identifying a correct image without false positives (seeing what isn't there) or false negatives (missing what is there).

The problems include poor image resolution (the camera may be too distant, or may be low-quality); blurry images, particularly if the vehicle is in motion; deliberate or inadvertent obscuring of the plate, such as by a tow bar or dirt; and an untypical style of lettering, such as italics.

The biggest problem is poor lighting. This can result from shadows cast by trees or buildings, poor weather, or darkness. It can also be the wrong type of light, or wrongly angled: solar glare, a low sun, or reflections from headlights or overhead lighting.

To overcome such problems, some countries now require counter-reflective plates, which bounce light back to its source. This improves the contrast between the numbers and letters and their surroundings. There are also stylistic limitations in force in many countries. Another approach is to alter cameras to improve the lighting available for them.

A pioneer in that kind of work is PIPS Technology, which is based in Hampshire. They have over 25 inventions including some under their old name Pearpoint. Their website claims that they believe themselves to be the biggest company in the world specialising in the field.

Their work includes the use of infrared light to improve the image. One of their patents, shown here, works by the camera suppressing glare and providing its own lighting.

Infrared illumination is shone onto the 'indicia' (number plates) which reflects it back. The camera can read a plate up to 22 metres away on a vehicle moving at up to 38 metres per second. A ring of infrared light emitting diodes (LEDs) surrounds the camera. To prevent blurring of the reflected indicia image, the infrared illumination is pulsed, with the camera shuttering synchronised to the pulses. An 'indicia image signal' is then processed. All this means that the reading apparatus is simple to use, easy to maintain and relatively cheap.

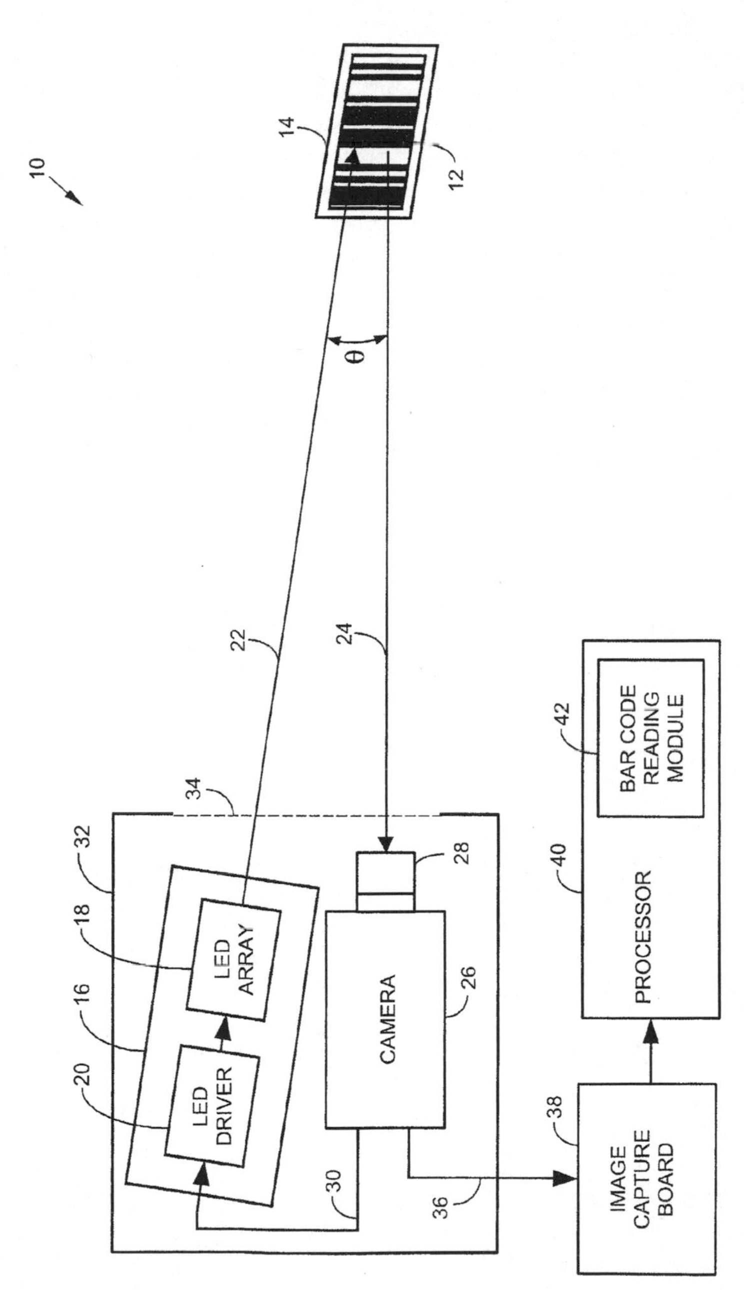

Apparatus for reading indicia from a remote location, EP 1246118, published 2 October 2002

The use of infrared light has the advantage of high reflectivity from the plates; it is invisible to the human eye, even at night; it filters out visible light, hence eliminating glare; and it penetrates rain and fog better than visible light.

The same patent proposes an ingenious use of infrared technology. Number plates could be manufactured so that they would display a covert bar code inside a reflective medium. This would be at the back of the frame that goes round the actual number which would be made of a material such as, they suggest, ICI Acrylic's Perspex Black 962, which allows infrared light to pass through, but which is opaque to visible light.

If infrared light is shone on the front of the number plate then the concealed bar code becomes visible. As the patent explains, 'This covertness thwarts those who might otherwise counterfeit a visible bar code for use on a vehicle to gain unauthorized access to a parking facility or for use on a toll road'. To add to the security, a visible but fake bar code could be placed on the front of the number plate.

It is interesting to know that the company is highly profitable. PIPS Technology in 2008 had a turnover of £12 million and an operating profit of £4 million. It's an ill wind . . .

Bricks from ash, not straw

Waste disposal is a perennial problem. It is therefore an excellent development when a waste product can be recycled into a new and useful one.

Henry Liu of Missouri has done just that for 'fly ash', the fine waste removed from the smoke emitted by coal-fuelled power stations. Only a third of the ash waste is at present reused, meaning that in the USA alone 40 million tons is disposed of annually in slurry pits or in mine workings. Turning that ash into some of the nine billion bricks the USA makes every year would both preserve supplies of the materials conventionally used to produce them and reduce the harmful byproducts of that process, which include pollutants such as mercury.

Professor Liu is a retired civil engineer. During his career he used hydraulic presses to compress goods for transportation, so making them cheaper to move. One such project, in 1999, involved the compression of fly ash.

He tried putting the white powder, mixed with water, through a hydraulic press to see what would happen at 28 megapascals of pressure – 10 bar or 145 lb per square inch, which is a lot of pressure. Within two weeks the mixture had set into blocks as strong as concrete. This made sense, as concrete sticks together because of its cement content, the calcium oxide in which binds with surrounding materials such as crushed rock when it reacts with water. Pure fly ash has calcium oxide levels of between 20 and 30%.

Liu spent eight years searching for the best method of making bricks from ash, helped by $600,000 of grant aid from the National Science Foundation. Much of the effort and money was needed to ensure that he met the mandatory American Society for Testing and Materials standard for bricks. Under this standard bricks are required to survive 50 cycles of freezing and thawing. Liu's cracked after just eight. He tried changing the shape and adding nylon fibres, but nothing seemed to work.

Then he tried adding a chemical which acts as an air-entrainment agent. It produces millions of microscopic bubbles in the hardened block, giving water less room to sneak in and extending the lifetime of fly-ash bricks to more than 100 freeze–thaw cycles. Another improvement was that his bricks required heating for only one day in a 65°C steam bath. That's hardly even warm in comparison to conventional clay bricks, which are heated to over 1,000°C.

As the bricks are made using mainly pressure rather than heat, less energy is required to produce them. Fly-ash bricks costs 20% less to make than conventional bricks, and are just as good. They are also more uniform because they are made using a mould.

In June 2005 Liu applied for WO 07/005065. Its initial pages describe in considerable detail the problems involved in improving the product, and previous research.

The invention has won awards and Liu was in the process of recruiting licensees to manufacture what he called a 'green brick'. 'The people who buy bricks will definitely be interested,' says Pat Schaefer, a sales manager for US manufacturer Midwest Block & Brick. 'But I don't see the brick companies liking it at all.'

Sadly, Liu was killed in a car accident in December 2009. Carla Roberts, Liu's administrative assistant at Freight Pipeline Company, has stated, 'We plan to continue the projects.'

Near-field communication – 'wave and pay'

I was returning home after making a presentation to the East London Inventors Club when I picked up a copy of the free newspaper, *The Wharf*, in a Docklands railway carriage. There was an article about a trial 'wave and pay' scheme at Canary Wharf using debit cards.

This is just one example of the concept that the experts call 'near-field communication' (NFC). In this case a cashless payment is made using a card which merely needs to be in proximity to the till rather than inserted into a card reader. But there is potentially much more to the concept than simply paying for the weekly shop.

Applications may be divided into passive and active modes.

In passive communication mode the 'initiator device' (the till) provides a carrier field and the 'target device' (the debit or credit card) answers by modulating the existing field. In this mode, the target device may draw its operating power from the initiator-provided electromagnetic field. This is the type of technology that would be used to process card transactions, though its use would not extend much beyond that.

The active communication mode has greater scope. In this mode both initiator and target device communicate by generating their own electromagnetic fields. Consequently, both devices typically require a power supply. One obvious target device suitable for active communication is the mobile phone. It is an object that most people carry, it is assigned to a particular person (who is therefore accountable), and it is battery powered. This would extend further the existing trend of turning mobile phones into multifunctional tools. They already double as cameras and mini computers, so why not as a mini bank, too?

A second advantage that mobile phones have over, say, smart cards is they possess keypads and display screens. PIN numbers could continue to be used, but entered on the phone's keypad rather than into a dedicated console, which would enhance both security and customer confidence. The screen would be used to receive or to confirm information during the transaction.

Mobile phones are key to many of the NFC concepts currently being researched. These might include using them to gain access to cars, buildings, or other secure areas, as electronic money, as tickets, and as travel cards. Data that could be read by phones could even be included on medicine containers and food packaging.

The potential uses of NFC are considered so important that the GSM Association has been set up to foster cooperation among mobile phone companies. Without uniform standards and protocols the parties to a transaction may not be able to communicate. Similar cooperation was needed in the case of fax machines and mobile phones themselves, and, for example, for television (a standard for which dates back to 1941 in the USA). In this case it is International Standard 18092.

There are so many patent applications that it is difficult to pick one as representative, but the illustration shown from a Philips invention was chosen because, to be frank, it amused me.

The idea behind the invention is that the mobile phone is used to authorise the purchase

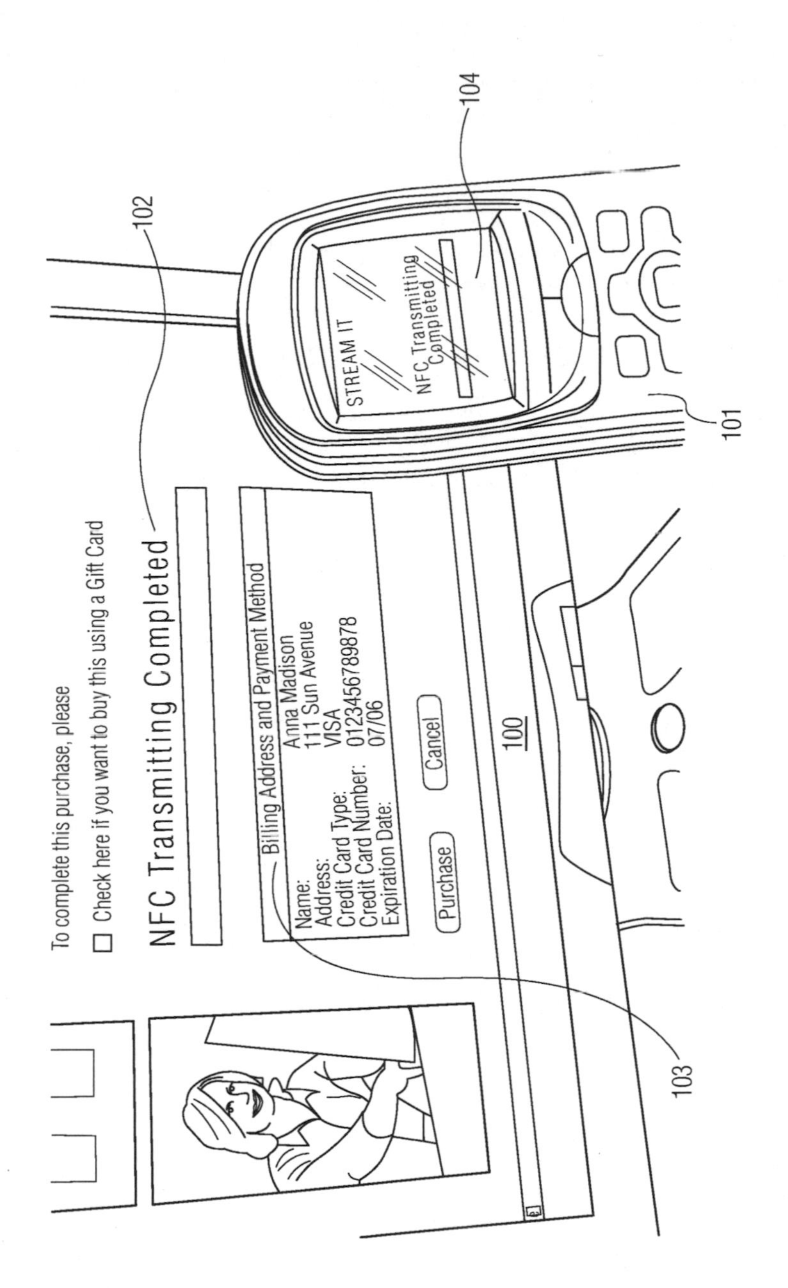

Ordering content by mobile phone to be played on consumer devices, WO 2006/077526, published 27 July 2006

of copyrighted content such as music or films. In return the user is authorised to use the content on DVD players or personal computers, for example.

I looked at this patent specification's appended search report, and it mentioned as X documents (potential killers of novelty) two web pages from news-gathering sites and another world application. The invention may not be new, but at least I found its drawings entertaining.

Incidentally, the article in *The Wharf* concluded by saying that some clubbers in America have had chips embedded in their bodies so that they don't have to carry cash with them when they go out. Sounds sensible, just so long as they don't find a cabbie who only accepts cash or cards . . .

Computer-generated imagery in films

I was watching a programme about the hundred best children's films when it was stated that *Monsters, Inc.*, which dates from 2001, was the first animation to include realistic hair and fur. It's the one where monsters are frightened by a little girl named Boo.

That movie was by Pixar. Out of curiosity I had a look and found that they have over 140 patent documents for algorithms to manipulate computer-generated imagery (CGI) to produce useful and interesting effects in animation. CGI dates back to *Westworld* in 1973, but the first film that could be described as using sophisticated techniques was *Jurassic Park* in 1993. Pixar have been filing patents in the field since 2003. They explain ever more sophisticated software to create realistic effects.

The first to be published is illustrated here. This shows the order in which the software carries out its functions. The specification's initial page discusses the problems of creating realistic effects and then goes on to explain the invention with language such as

> The computer program product is on a tangible media that includes code that directs the processor to determine a grid including a plurality of voxels and a plurality of vertices, wherein at least a subset of the voxels bound at least a subset of the hair objects, code that directs the processor to determine a hair density value associated with each of at least a first set of vertices, and code that directs the processor to filter the hair density value associated with each vertex from at least the first set of vertices with a filter to form filtered density values.

Whew.

Pixar also have a later 'Volumetric hair rendering' patent, which says, 'Manually animating each individual hair is typically too time-consuming to be practical. As a result, the motion of hair is often derived from a physics-based simulation. Hair simulations typically represent each hair as a set of masses and spring forces' – such as gravity or wind. Yet that can look false, as hair does not move in a single mass. There is also the problem of appropriate lighting. The trick was to get hair to move realistically in a 'collective' fashion without an expensive amount of computing time. The patent's solution was 'volumetric representation [which] determines hair responses to hair to hair collisions and illumination'.

Also active in CGI is DreamWorks, who have published US 2007/0270092 on how to create a realistic look for feathers, something they point out was not achieved in early animation. They say that feathered characters were traditionally rare for this reason. Even with more modern animation there was a problem when feathers would inconsistently 'pop' from frame to frame because of the way the feathers were layered. The basic idea in the invention was to avoid intersections of the 'feather-proxy' elements.

They also have the intriguingly named 'Fast oceans at near infinite resolution', US 2009/0207176. This involves using three-dimensional wire-frame models, which are a mesh-like look on which animation can be placed. Wire frames of this type are simple and fast for the calculations required to ensure smoothly working animation, especially in fast-moving action.

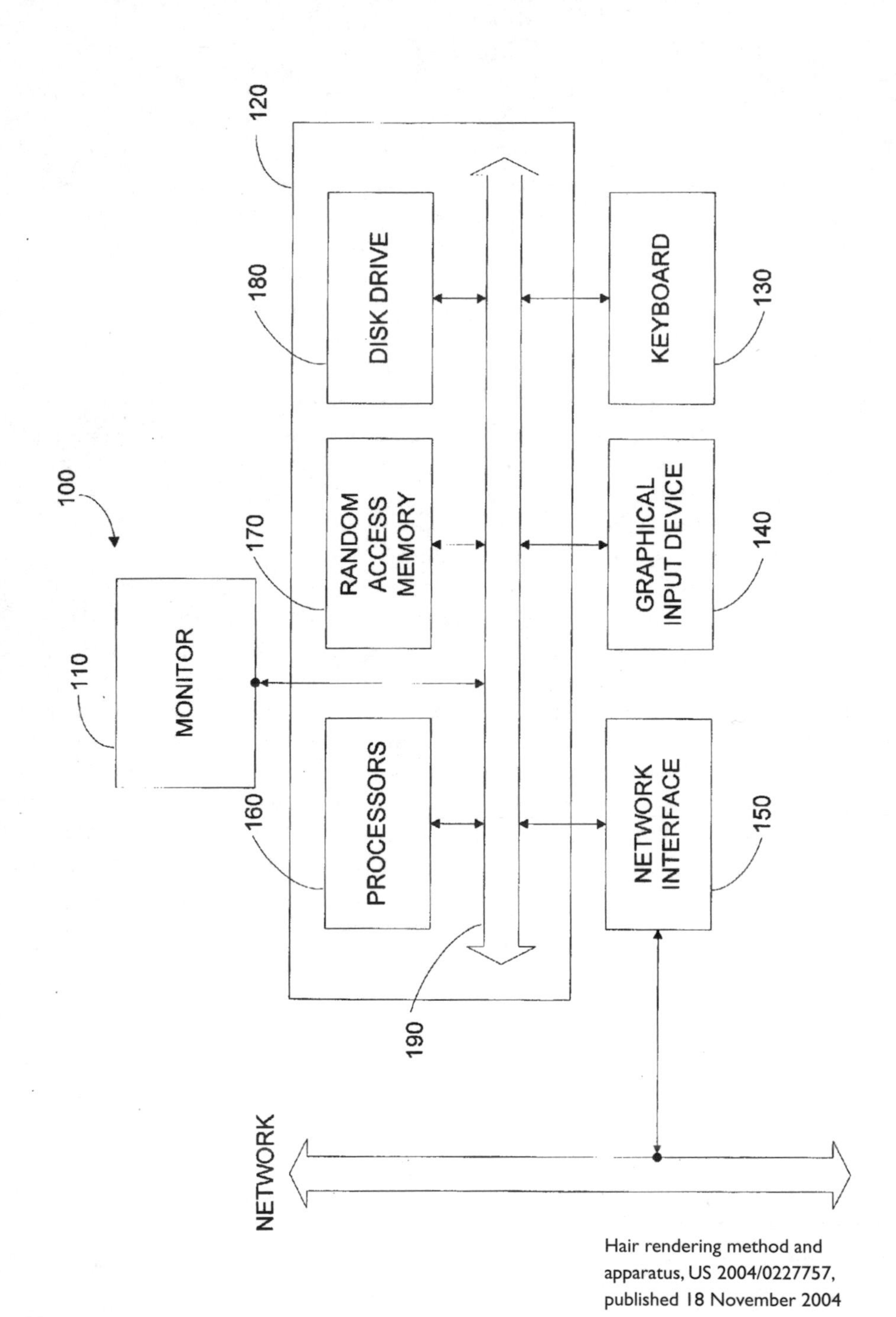

Hair rendering method and apparatus, US 2004/0227757, published 18 November 2004

This patent application talks of animating a body of water by 'deconstructing' a master wave model into three models for different kinds of waves. An optimised wave model is then built up using them, and specific wave heights are associated with specific points.

Disney is another major player, and they have been responsible for the patent application illustrated below, where rather than supplying special glasses to give a 3-D effect the image itself is manipulated to give that effect inside the completed film.

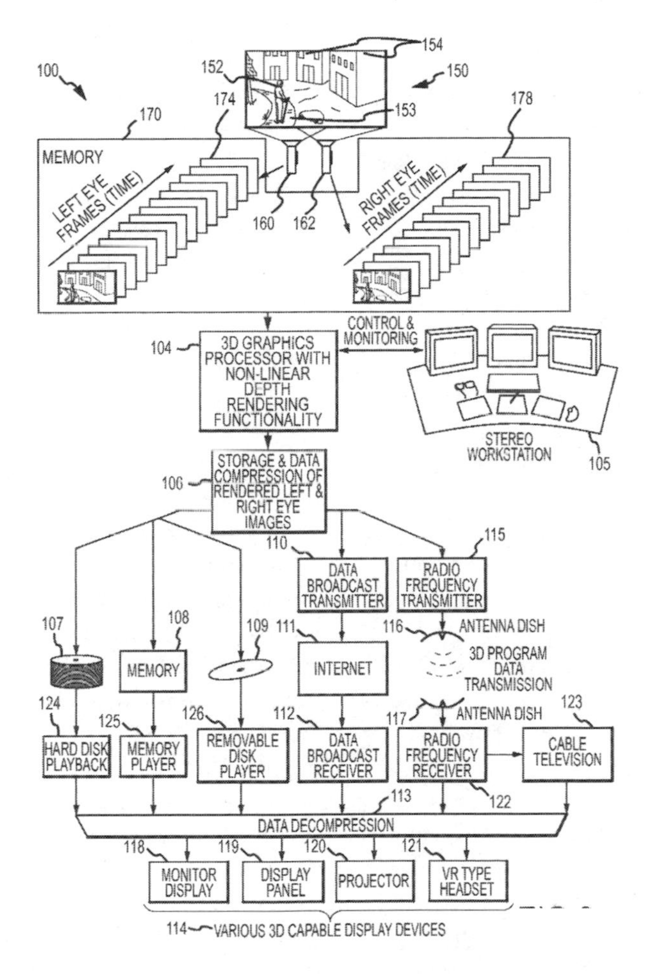

Non-linear depth rendering of stereoscopic animated images, WO 2009/111150, published 11 September 2009

The future for the car?

The internal combustion engine has ruled supreme for a century. That it would do so was not obvious at first. Electric vehicles outsold petrol- (and steam-) driven cars in the USA in 1900. But they were only the biggest fish in a tiny pond. Electric-vehicle production peaked in 1912 and then faded away. They were more expensive to make and to run than their rivals and the cheaper option prevailed.

Today, however, increasing concerns about pollution, rising fuel prices and diminishing oil reserves have stimulated innovation in the field. Legislative pressures such as stricter emission controls and performance limits have also played a part. Many different (and, of course, incompatible) approaches are being investigated and manufactured. In the five-year period 1989–93 there were under 1,700 inventions in the electric vehicles class in the priced Derwent World Patent Index database. This number had climbed to 28,000 in 2004–8. About 70% of those are from Japan, Toyota leading with 5,282, followed by Nissan and Honda. General Motors had just 289. It is very rare that it is possible to attribute patents to a specific model, especially as many are never incorporated into a commercially produced vehicle.

The production of electric vehicles is beset with problems. They are engineered almost entirely differently from petrol-driven ones; battery power alone is rarely sufficient to sustain one for more than about 150 kilometres; they are weighty and bulky; and they are very expensive. Some models can be recharged at home, others need to find one of the relatively few special recharging points available at filling stations. Work is also proceeding on the feasibility of a speedy remove-and-replace system so that exhausted batteries could be renewed in a few minutes, instead of the driver spending an hour charging up.

Many models have been launched, mostly with very few sales. The best known electric car in Britain is the tiny Smart car, designed in the 1990s to a specification that stipulated that it must be capable of being parked end-on to the kerb. It is 2.5 metres long, and the standard engine size is just 698cc. The original, purely electric concept has blurred with models available with petrol turbo, diesel turbo and so on.

The technology in those cars belongs to Daimler, and Daimler Chrysler Corporation has been responsible for the first-illustrated patent. It simulates the feel of an internal combustion engine when coasting; braking is controlled when going downhill; and the energy usually lost in braking is converted to stored electric energy.

Hybrid approaches combining electric and conventional engines are being increasingly advocated and marketed. A small internal combustion engine is combined with one or more electric motors in any of a number of variations on the basic concept. Using the same idea as in the previously described invention, energy used in braking (while using the petrol-driven engine) is frequently used to charge the batteries. There are also savings when moving off from a halt, with the electric motor taking over, the petrol engine having been switched off. These, and the in-motion recharging of the batteries by the petrol engine rather than from the mains, are the main advantages of the hybrids over the purely electric variant.

The Toyota Prius, launched in 1997, was the first widely available hybrid vehicle. Two million hybrids have been sold by the company since then, mainly of that model. It is a 'pure' hybrid, in

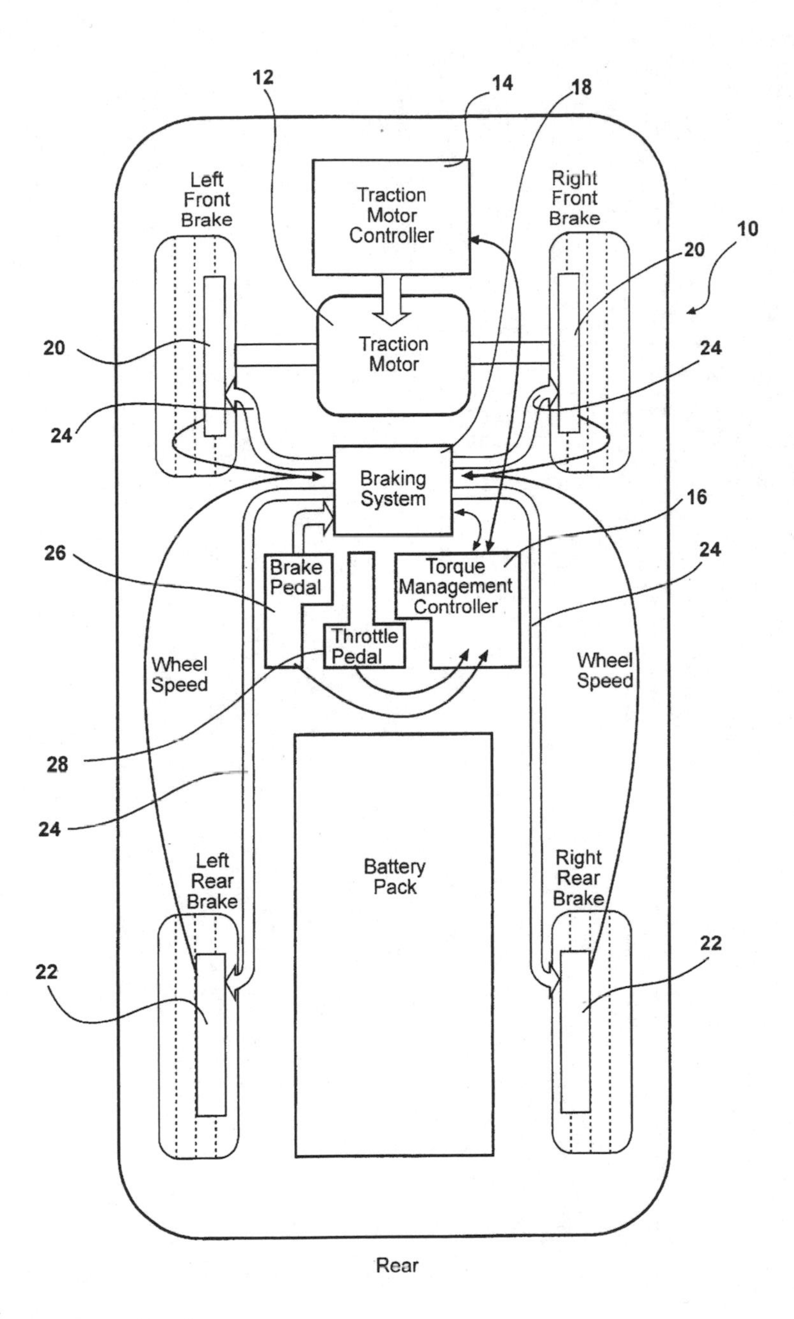

Intelligent coast-down algorithm for electric vehicle, US 6364434, published 2 April 2002

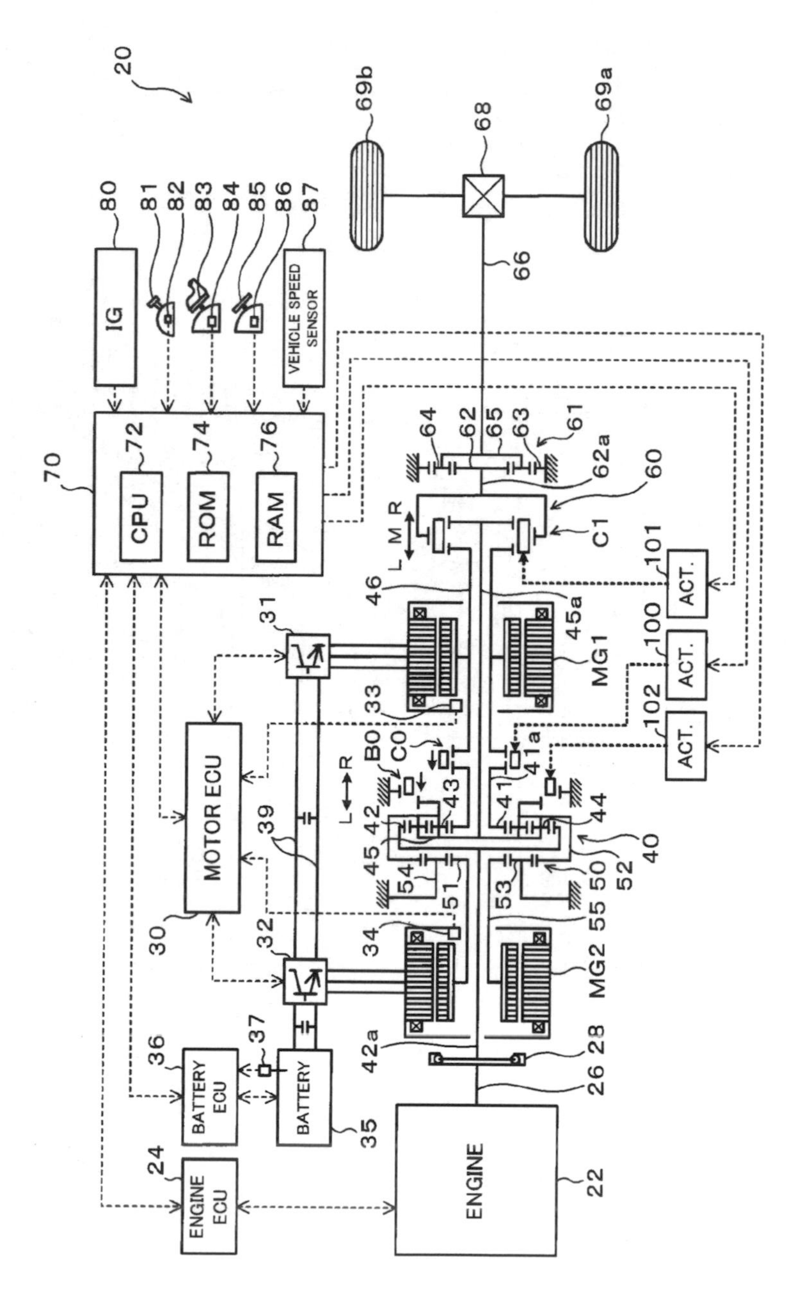

Power output apparatus and hybrid vehicle, US 2010/012407, published 21 January 2010

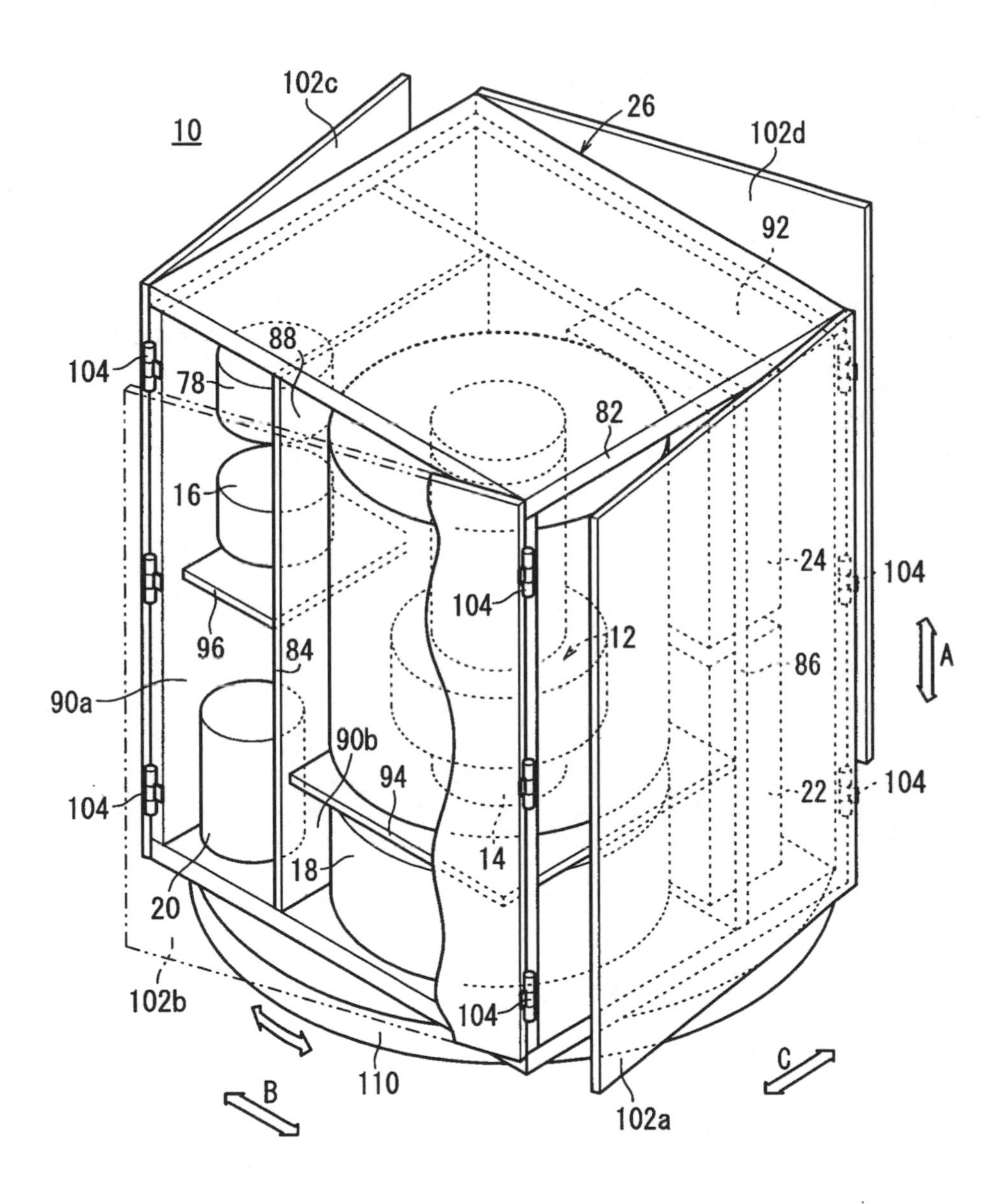

Fuel cell system, WO 2008/114570, published 25 September 2008

that the electric motor alone can be used to drive the car for some distance.

Many inventions in the field involve the software, electronic, and mechanical aspects joining together. An example of a 'schematic configuration view', also by Toyota, is illustrated on p.146.

Its summary explains that it contains a 'transmission differential rotation mechanism that is a single pinion planetary gear mechanism having a sun gear, a ring gear, and a carrier that is connected to a drive shaft', which differentially rotate with each other. The clutch can selectively couple 'one of or both of the carrier and the sun gear of the power distribution and integration mechanism with the sun gear of the transmission differential rotation mechanism'. It is for a rear-wheel-drive engine where the petrol engine is (22), the electric motors are MG1 and MG2, ECU is the electronic control unit and CPU is the central processing unit.

Another proposed innovation is to use hydrogen as a fuel. This could be stored in the vehicle under very cold, pressurised conditions. Otherwise it could be generated from feedstock (a raw material such as gas) and then used either in an internal combustion engine, or in fuel cells. An electrochemical reaction (using oxygen as well as hydrogen) creates energy. Water is a by-product, with some warmth given off as well.

The best known example of a car that uses hydrogen in fuel cells is the Honda Clarity, which became available in 2008. At present American residents are limited to leasing one for $600 per month in the Los Angeles area, where there are 16 hydrogen filling stations.

Honda has over 200 patent specifications for fuel-cell technology. One is illustrated on p.147.

Most of Honda's inventions in the field involve a hydrogen supply carried aboard the vehicle, but this one includes a reformer, where hydrogen is extracted from an onboard feedstock. The idea behind the invention was to rearrange the components so that the low-temperature areas were not affected by areas emitting heat. The fuel cell module is at (12) while (18) supplies oxygen from the air. The fuel gas is supplied from (16), (22) is for power conversion, and (24) is for the control apparatus.

It is likely that only a few of the inventions in the field, let alone types of vehicle, will survive to be important forerunners of technologies that will be in use a few decades from now.

How the patent system works, and suggestions for research

Anyone interested in researching a particular subject, company or inventor will benefit from understanding some basic principles of the patent system that are generally accepted worldwide. What follows is, of necessity, only a simplified explanation.

Patents are for functional aspects of technology and for many products are accompanied by designs for the look of things, and by trade marks for names or logos for products or services, where the familiar ® (if registered) or ™ (if unregistered) is used. Together they are called 'industrial property', and can be registered at patent offices for protection in those countries. If copyright for literary, artistic and musical works (and this can include websites and software) is added then all these varieties are included in the term 'intellectual property'.

Industrial property as an internationally understood system dates back to the Paris Convention of 1883. It enables those asking for protection abroad to quote their original 'priority' filing details when applying at foreign patent offices, which must be within 12 months of the original filing date. Priority means that, for example, an applicant who files in China before another applicant files in Germany would have 'priority' worldwide, and would therefore be granted a patent for that invention, even if neither applicant intended to patent in the other's, or the same, country. Only one country rejects this principle. It is the United States, where the principle is 'first to invent', and proof of priority has to be provided in disputes between rival applicants.

Most countries publish the application 18 months from priority and later publish (if accepted as new) a separate grant document, setting out the protected claims.

These documents consist of descriptions of the inventions, drawings or diagrams if relevant, and claims setting out the monopoly requested or granted. Cover pages containing address, names, dates and other data plus a summary are common, and there may be a search report setting out known prior art. I tend to talk of patent specifications or documents as the term 'patent' means that one has been granted, which may not yet be true.

It is very difficult to write a patent application without the aid of a patent attorney, who also argues the case for granting the patent with the patent office if there is any dispute.

Protection for granted patents generally lasts for 20 years from filing, but it often lapses if renewal fees are not paid at intervals. After that period the patent expires and anyone can use it. Litigation can be pursued to revoke a patent if it can be shown that prior art existed. If no patent is granted in a specific country then there is no protection in that country. The principle is that the protection is available in exchange for disclosing details of how the invention works, and in theory should help small companies benefit from their innovations. Large companies would otherwise find it easy to copy others' work.

Many of the companies and independent inventors described in this book used the 'World' (Patent Cooperation Treaty) system which publishes a single application (although each country decides separately whether or not to grant). There is also the European system, which results in a single patent being granted for most of Europe (but they can lose protection separately within each country, depending on litigation or renewal fees).

Each patent specification describing the invention has a number and often a document code to identify it. Two-letter codes are used as prefixes such as WO for World, GB for the United Kingdom, US for the United States and so on. Usually 'A' denotes the initial published application and 'B' the granted patent. The number may be the same for the A and the B, as for Britain, currently publishing in a 2 million number range. Sometimes they are different, as in the USA. There the applications begin with a year prefix and the Bs are in a series currently in the 7 million number range. These are called utility patents, as opposed to design patents (which are prefixed by a D and are currently in a 600,000 number range).

With a few exceptions, such as South Africa and Argentina, most countries with a sizable industrial sector publish their patent documents on the Internet as digital files, commonly in PDF format, often within a few weeks of publication, or even on the day itself. An increasing number of backfiles are available for many countries. Over 50 million patent documents are available in this way.

This number may sound high but there are far fewer inventions than documents: some duplicate earlier patents and are rejected, there are often A and B formats for the same invention, and 'equivalents' in foreign countries of the original patent application form 'patent families'. They can be accessed by patent number and sometimes by title, classification, name and so on. The titles are often searchable in English even if the patents are in other languages, and free machine translations are available for many Japanese patents.

All this means that patent specifications form a highly organised literature which can be searched to find material of interest. Very often the problem is not that of finding anything relevant, but of sorting through huge amounts of tenuously connected hits to find the gems.

There are a number of useful and free websites containing patents, where, for example, patents mentioned in this book, or companies or subjects, can be searched for. It is very easy to overlook relevant material when conducting a detailed search online, especially one to determine whether an invention is new. It is better to ask for help at patent libraries open to the public. Their staff can also explain how the patent system works in more detail than I am able to provide here.

The patent libraries in the United Kingdom are listed at www.epo.org/patents/patent-information/patlib/directory/unitedkingdom.html, while those for the USA are listed at www.uspto.gov/products/library/ptdl/. In other countries, apply to the national patent office for help.

The key free patent online database is esp@cenet at ep.espacenet.com/. It contains a vast number of patents as PDFs, including those published in the USA back to 1836 and in Britain back to 1895. It can be confusing to use and number formats for patents can vary. In order to get the best from it, new users should read the tutorial provided at 'Get Assistance' in its 'Advanced Search' option. Most of the patent numbers cited in this book can be found on it in formats such as US2009309882 or WO2009079514. Sometimes inserting or dropping a zero helps to locate a document.

The same database also contains the ECLA patent classification which can be explored to identify classes of interest, or to narrow the search area and avoid irrelevant material. Despite the name, ECLA does not cover Europe alone, but is a frequently more detailed variant of the universally used International Patent Classification, which is printed on patent documents. They can be selected for searching by clicking the box next to the class, then Copy. The ECLAs generally cover patent specifications from the USA, European countries, and the World system (back to 1920 at least), but not Far Eastern patents.

Other useful databases include Free Patents Online at www.freepatentsonline.com, where the most relevant hits are ranked at the beginning of the results, and Google Patents at www.google.com/patents, which includes design patents. Both allow searching of the complete text, unlike esp@cenet.

An ordinary Google search for inventions will often find recent American patents from a number of free sites if the word 'patent' is included in the search terms in esp@cenet. This is useful but should not be relied on exclusively.

Finally, I hope you will find it helpful to browse through my patent search blog at britishlibrary.typepad.co.uk/patentsblog/.

Further reading

Copies of all, or nearly all, these books are held in the British Library, often in the Business & IP Centre. The more sober textbooks meant for the professional have largely been omitted, as anyone intending to file for a patent is advised to use professional help. Many reflect a US viewpoint, where the laws as well as the business culture can be somewhat different to the UK.

Inspiration from the past

Century Makers: one hundred clever things that we take for granted which have changed our lives over the last one hundred years. D. Hillman and D. Gibbs. London: Weidenfeld & Nicolson, 1998.

Historical First Patents: the first United States patent for many everyday things. T. Brown. Metuchen, NJ: Scarecrow Press, 1994.

Ingenious Women. D. Jaffé. Stroud: Sutton Publishing, 2003.

Inventing the American Dream: a history of curious, extraordinary and just plain useful patents. S. van Dulken. London: British Library, 2004.

Inventing the 19th Century: 100 inventions that shaped the world. S. van Dulken. London: British Library, 2001.

Why Didn't I Think Of That? 101 inventions that changed the world by hardly trying. A. Rubino, Jr. Avon, MA: Adams Media, 2010.

Humour

Absolutely Mad Inventions. A.E. Brown and H.A. Jeffcott, Jr. New York: Viking Press, 1932 [Reprinted 1970, 2001].

American Sex Machines: the hidden history of sex in the U.S. Patent Office. H. Levins. Holbrook, MA: Adams Media, 1996.

Great British Inventions. M. Tanner. London: Fourth Estate, 1997.

Inventions Necessity is Not the Mother of: patents ridiculous and sublime. S.V. Jones. New York: Quadrangle, 1973.

Patent Nonsense: a catalogue of inventions that failed to change the world. C. Anderson and I. Brown. London: Michael Joseph, 1994.

Patently Absurd: the most ridiculous devices ever invented. C. Cooper. London: Robson Books, 2004.

Patently Silly: the daftest inventions ever devised. D. Wright. London: Prion, 2008.

Totally Absurd Inventions: America's goofiest patents. T. Vancleave. Kansas City, MO: Andrews McMeel, 2001.

The World's Stupidest Inventions. A. Hart-Davis. London: Michael O'Mara, 2003.

Patent searching and documentation

British Patents of Invention 1617–1977: a guide for researchers. S. van Dulken. London: British Library, 1999.

Introduction to Patents Information. S. van Dulken (ed.). London: British Library, 2002.

Patent Searching Made Easy (4th edn). D. Hitchcock. Berkeley, CA: Nolo, 2007.

Innovation and commercialisation

A Better Mousetrap: the business of invention. G. Barker and P. Bissell. Hebden Bridge: www.abettermousetrap.co.uk, 2007.

From Patent to Profit: secrets and strategies for the successful inventor (3rd edn). B. DeMatteis. Garden City Park, NY: Square One, 2004.

How to Make Patent Drawings Yourself (3rd edn). J. Lo and D. Pressman. Berkeley, CA: Nolo, 2001.

Inside the Patent Factory: the essential reference for effective and efficient management of patent creation. D. O'Connell. Chichester: Wiley, 2008.

Inventing for Dummies (UK edn). P. Jackson, P. Robinson and P.R. Bird. Chichester: Wiley, 2008.

Inventive Thinking Through TRIZ: a practical guide. M.A. Orloff. Berlin: Springer, 2003.

The Inventor's and Innovator's Kick-start Guide. J. May. Montrose: Two Little Ducks, 2005.

iProperty: profiting from ideas in an age of global innovation. W.A. Barrett, C.H. Price and T.E. Hunt. Hoboken, NJ: Wiley, 2008.

Patent It Yourself (13th edn). D. Pressman and S. Elias. Berkeley, CA: Nolo, 2008.

The Patent Writer: how to write successful patent applications. B. DeMatteis, A. Gibbs and M. Neustel. Garden City Park, NY: Square One, 2006.

Patents, Registered Designs, Trade Marks and Copyright for Dummies. J. Grant, C. Ashworth and H.A. Charmasson. Chichester: Wiley, 2008.

Patents, Trademarks, Copyrights, and Trade Secrets: what automation professionals, manufacturers, and business owners need to know. W. Buskop. Research Triangle Park, NC: ISA, 2008.

A Practical Guide to Drafting Patents. G. Roberts. London: Sweet & Maxwell, 2007.

Standards and Intellectual Property Rights: a practical guide for innovative business. M. Clarke. London: British Standards Institution, 2004.

The Toy and Game Inventor's Handbook. R.C. Levy and R.O. Weingartner. New York: Alpha, 2003.

Index

Acknowledgements

I would like to thank Lara Speicher, who commissioned this book, Jenny Lawson, my editor, Trevor Horwood, my copy-editor, and Mercer Design, who designed it.

First published in 2010 by
The British Library
96 Euston Road
London NW1 2DB

British Library Cataloguing in Publication Data
A catalogue record for this publication is available from The British Library

ISBN 978-0-7123-5802-6

Designed by Mercer Design, London
Typeset by Hope Services (Abingdon) Ltd
Printed and bound in Great Britain by MPG Books Ltd